SpringerBriefs in Crystallography

SpringerBriefs in Crystallography, published under the auspices of the International Union of Crystallography, aims at presenting highly relevant, concise monographs with an intermediate scope between a topical review and a full monograph. Areas of interest include chemical crystallography, crystal engineering, crystallography of materials (ceramics, metals, organometallics, functional materials), instrumentation, mathematical crystallography, mineralogical crystallography, physical properties of crystals, structural biology and related fields.

SpringerBriefs present succinct summaries of cutting-edge research and practical applications covering a range of content from professional to academic and featuring compact volumes of 50 to 125 pages.

- A timely report of state-of-the art experimental techniques and instrumentation
- New computation algorithms or theoretical approaches
- A bridge between new research results, as published in journal articles, and a contextual literature review
- A snapshot of a hot or emerging topic
- An in-depth case study
- A presentation of core concepts that students must understand in order to make independent contributions

Briefs are characterized by fast, global electronic dissemination, standard publishing contracts, standardized manuscript preparation and formatting guidelines, and expedited production schedules.

Publications in these series help support the Outreach and Education Fund for the International Union of Crystallography.

John R. Helliwell

Certifying Central Facility Beamlines for Biological and Chemical Crystallography and Allied Methods

John R. Helliwell
Department of Chemistry
University of Manchester
Manchester, UK

ISSN 2524-8596 ISSN 2524-860X (electronic)
SpringerBriefs in Crystallography
ISBN 978-3-031-80183-9 ISBN 978-3-031-80181-5 (eBook)
https://doi.org/10.1007/978-3-031-80181-5

This Springer imprint is published by the registered company Springer Nature Switzerland AG
The registered company address is: Gewerbestrasse 11, 6330 Cham, Switzerland

If disposing of this product, please recycle the paper.

Preface

My aims in writing this Springer Brief book are:—

Firstly, to guide beamline providers and users on what to look for in selecting experiments for a given type of facility and beamline so that beamtime usage and effectiveness are maximised. Secondly, to help them to navigate facility publication, as well as data management and sharing policies, and avoid potential pitfalls between facility and user on these important post experiment aspects. Thirdly I provide tabulations of these experimental facilities globally. Often in my text I express my personal perspectives from my career these past 50 years. That said I am very grateful to the six colleagues who gave me their expert comments, who I list below.

I need to air what is a Springer Brief book? These are defined to be 'succinct summaries of cutting-edge research and practical applications covering a range of content from professional to academic and featuring compact volumes from between 50 to 125 pages'. The aims (https://www.springer.com/series/16236) include such as provide a 'bridge between new research results and a contextual literature review' and, rather more challenging, 'A presentation of core concepts that students must understand in order to make independent contributions'. In general, the word 'brief' I think has several different meanings, namely: short duration, concise summary or inform someone thoroughly especially in preparation for a task. The first of these meanings relates to these books being relatively quick to read and hopefully digest. The second meaning is that they are not verbose and are a summary. The third is the more challenging aspect especially so with respect to students where practical experimental experience is as important, or more so actually, as reading a book or having tutorials.

Some 30 years ago I published my *Macromolecular Crystallography with Synchrotron Radiation* research monograph (Helliwell (1992)) which comprised more than 600 pages. At the time I was also heavily involved proposing, developing, and managing beamlines for macromolecular crystallography at the UK's SRS. My efforts included generating spin off applications into synchrotron radiation chemical crystallography and solution X-ray scattering. I helped launch analytical services to industry, which became the Daresbury Analytical Research Technical Services (known simply as DARTS). The beamlines operating at the UK's SRS provided a platform for the proposals for macromolecular crystallography at the European Synchrotron Radiation Project, based at that time in the mid-1980s at CERN in Geneva. These proposals formed a part of the ESRF Foundation Phase Report (1987), which led to the construction of the ESRF as a facility in Grenoble in the 1990s.

My monograph book was published in paperback in 2005. I considered a second edition, but the field had expanded so enormously and had helped to stimulate neutron Laue macromolecular crystallography at the Institut Laue Langevin also in Grenoble, which needed a description alongside the synchrotron applications. Overall, when the opportunity came up to write a Springer Brief book in 2023, I thought 'aha' this is a good opportunity to bring the very practical aspects of using beamlines up to date. I hope this Brief book will be as useful as my synchrotron radiation monograph has obviously been, having made it into paperback.

In this Brief Book I have added electron diffraction and electron bioimaging as these are now also operated by the synchrotron radiation facilities. To form as complete a description of the landscape of methods, I have included solution scattering and several spectroscopic methods.

Manchester, UK   John R. Helliwell
October 2024

Acknowledgements

I am grateful to Prof. Dr. Massimo Nespolo for his help and constructive criticism in formulating my Springer Brief proposal and to Emeritus Professor Claude Lecomte for his constructive comments. I dedicate a heartfelt thanks to all my students and research staff, in the universities and in the scientific civil service, and to the many colleagues I have met in conferences and in my representational and community service roles, who have all greatly enriched my interests and my understanding, which I hope shine through my book as practical experience. Specifically, the following of my past and current colleagues and Ph.D. students kindly read through and commented on my text draft: Andrew Thompson (formerly Director of Life Sciences at Soleil now Head of their cryoEM Centre, Paris, France), Michele Cianci (formerly EMBL Hamburg now Professor at the Università Politecnica delle Marche, Italy), Dr. Matthew Blakeley (Life Sciences Group, Institut Laue Langevin, Grenoble, France), Dr. Eddie Snell, Principal Research Scientist at the Hauptman-Woodward Medical Research Institute in Buffalo, USA, Dr. Alice Brink of the Chemistry Department, University of the Free State, Bloemfontein, South Africa, and Dr. Lucrezia Catapano researcher at the MRC Laboratory of Molecular Biology, Cambridge, UK. Any errors or misconceptions are of course my own.

I was greatly encouraged in considering writing this Springer Brief by my giving the opening lecture at the Synchrotron radiation Workshop entitled "*A practical approach to synchrotron experiments*", which was on day zero of the IUCr Congress in Melbourne Australia held in August 2023. This Workshop was splendidly organised by Dr. Dubravka Sisak-Jung under the auspices of the Swiss Crystallography Association.

Contents

About the Author

John R. Helliwell, D.Sc. (Physics, University of York, 1974), D.Phil. (Molecular Biophysics, Oxford University, 1978), is Emeritus Professor of Chemistry at The University of Manchester, where he served as Professor of Structural Chemistry from 1989 to 2012. Academic teaching from 1979 till 1988 was at the Universities of Keele and York in the physics departments there. He is a Researcher in the fields of crystallography, biophysics, structural biology, structural chemistry, and data science. He was also based at the Synchrotron Radiation Source at the UK's Daresbury Laboratory, in various periods of appointment between 1979 and 2008, including in 2002 as Director of Synchrotron Radiation Science. He is a Fellow of the Institute of Physics, the Royal Society of Chemistry, the Royal Society of Biology, and the American Crystallographic Association, and an Honorary Member of the British Crystallographic Association and of the British Biophysical Society. He is an Emeritus Member of the Biochemical Society. He is a Corresponding Member of the Royal Academy of Sciences and Arts of Barcelona, Spain, and an Honorary Member of the National Institute of Chemistry, Slovenia. His research awards include the European Crystallographic Association Eighth Max Perutz Prize 2015, the American Crystallographic Association Patterson Award 2014, and the 'Prof. K. Banerjee Endowment Lecture Silver Medal' of the Indian Association for the Cultivation of Science (IACS) 2001. He has published over 200 scientific research papers and several books, e.g., Macromolecular Crystallography with Synchrotron Radiation with Cambridge University Press (1992), published in paperback in 2005, as described above, and Macromolecular Crystallization and Crystal Perfection with N. E. Chayen and E. H. Snell), Oxford University Press—International Union of Crystallography Monographs on Crystallography (2010). He has published *Perspectives in Crystallography*, also now in paperback, to coincide with the International Year of Crystallography in 2014 as well as several Scientific Life, popular science, books in recent years, which are with CRC Press, Taylor and Francis.

He has served in roles of major responsibility such as President of the European Crystallographic Association (2007–2010), Chairman of the International Union of Crystallography's (IUCr) Commission on Journals (1996–2005), Chairman of the IUCr Diffraction Data Deposition Working Group (2011–2017) and its Committee on

Data (2017–2023) as well as its Representative to the International Council for Scientific and Technical Information (ICSTI; 2005–2014) and the International Council of Science's Committee on Data 'CODATA' (2012–2023). In the past thirty years, he has chaired several international advisory committees for synchrotron and, more recently, neutron, facilities' development, and their users' science. He was Leader of the UK Delegation at the International Union of Pure and Applied Biophysics Congress and General Assembly in New Delhi, India, in 1999 and was Leader of the UK Delegation at the International Union of Crystallography Congress and General Assembly in Prague in 2021.

In terms of research projects experience, this really had chance to expand in 1989 when he became Professor of structural chemistry at the University of Manchester. Initially his emphasis continued on beamlines' instrumentation, methods, and applications. Besides, synchrotron radiation similar themes emerged with neutrons as probe and studies of protein crystal perfection and crystal growth in microgravity. But several structural crystallography project themes steadily emerged mainly on structural biology of protein saccharide interactions, time-resolved catalysis studies of the enzyme hydroxymethylbilane synthase and the crustacyanin colouration protein which has bound carotenoids. These structural studies included approximately 100 PDB depositions. In terms of research skills, an emphasis on computational crystallography within these studies built on his software writing experience with Fortran during his early career. This theme lapsed as he saw, and extensively used, the better software from CCP4 and others such as Phenix and Shelx. Diffraction data processing is amongst the most effective for evaluating the quality of a beamline's performance, aspects covered in Chap. 26.

Chapter 1
A Beginner's Guide to Synchrotron Radiation and X-Ray Lasers

For the new user of a beamline, details such as its beam focus sizes and fluxes onto the sample will be of keen interest. Likewise, instrument scientists are keen to document their evidence for their beamlines' performances. The first such survey was in an article in SR News (Helliwell 1992) as part of the new IUCr Commission on Synchrotron Radiation activities. It was well received by the global user community as a useful benchmarking exercise. These numbers were not so easy to find as now via simple web searches where every beamline provides these details at their websites. In 1992 each beamline scientist had to be contacted directly! Thereafter the evolution of the sources and the improvements to beamlines had these benchmark performances to reference.

The recent major step change to extremely bright sources (EBS) was with MAX IV, and the high energy sources so that ESRF is now the ESRF EBS. Similar upgrades are also underway at the APS and planned for Spring8. These present an interesting variation within this theme i.e. a before and after level of performance. On a historical note, this was seen also at the SRS in the UK when a high brightness lattice was introduced in the mid-1980s which improved the intensities at the crystal sample by about 10 times, mainly due to a better focal spot size in the horizontal direction. The motivations behind the 1992 IUCr Commission on Synchrotron Radiation survey were several. Principally the bringing together of information which had resided primarily in printed Facility Annual Reports was brought into a single newsletter article. Thereby parameters for instrumentation could be compared to facilitate developments and improvements at all sources. It is difficult to document that was achieved but comments within the IUCr at meetings, and within the Commission on Synchrotron Radiation itself, were very favourable. The 1992 survey article (Helliwell 1992) concluded with the following three decades' forward look:

© The Author(s), under exclusive license to Springer Nature Switzerland AG 2025
J. R. Helliwell, *Certifying Central Facility Beamlines for Biological and Chemical Crystallography and Allied Methods*,
SpringerBriefs in Crystallography, https://doi.org/10.1007/978-3-031-80181-5_1

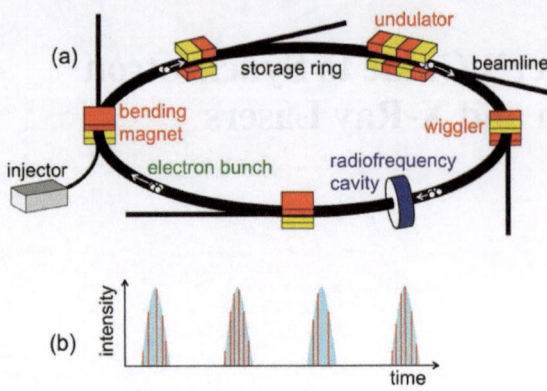

Fig. 1.1 a Generic scheme of a synchrotron radiation facility with its accelerator (storage ring), the electron injector, a radiofrequency cavity, and X-ray sources of different types with their beamlines. The electrons circulate in the ring as regularly spaced bunches. **b** Each time an electron bunch passes through a magnet, it emits a pulse of radiation, which includes micro pulses caused by individual electrons. Reproduced from Hwu and Margaritondo (2021) with the permission of IUCr Journals

The year 2020, then, promises a strong activity in the X-ray analysis of proteins and viruses supported with an increasingly large number of SR machines. The overall aim of this area of research is to achieve a huge database of such structures, a detailed knowledge of the catalytic properties of these molecules and a full understanding of biology at the molecular level such that it can be properly harnessed for medical applications including drug design. The challenge to physics is strong in terms of instrumentation and methods and to chemistry in terms of the knowledge and prediction of reactivity and interactions of one structure with another.

Today in 2024 these aims remain very central.

Now to describe the fundamentals. The basic components of an electron storage ring radiation source are shown in Fig. 1.1.

One of the first things a user encounters at the synchrotron ring is a machine and beamlines status. This is highly informative and self-evident. See the Soleil synchrotron display, a model of clarity, Fig. 1.2.

The ring consists of a number of dipole (bending) magnets to bend the electron beam along a circular trajectory to achieve a closed orbit. The bending magnets act as the basic sources of radiation. They are separated by straight sections in which are located numerous components including insertion devices. The straight sections house quadrupole, sextupole and octupole magnets to focus the electron beam. These are important components giving long term stability, particularly in beam position. The beam properties are determined largely by their arrangement, often referred to as the magnet lattice, which can vary greatly from one ring to another. Radio-frequency (r.f.) cavities are used to accelerate the beam to the required energy and to make up the energy lost due to the emission of synchrotron radiation; the r.f. cavities generate

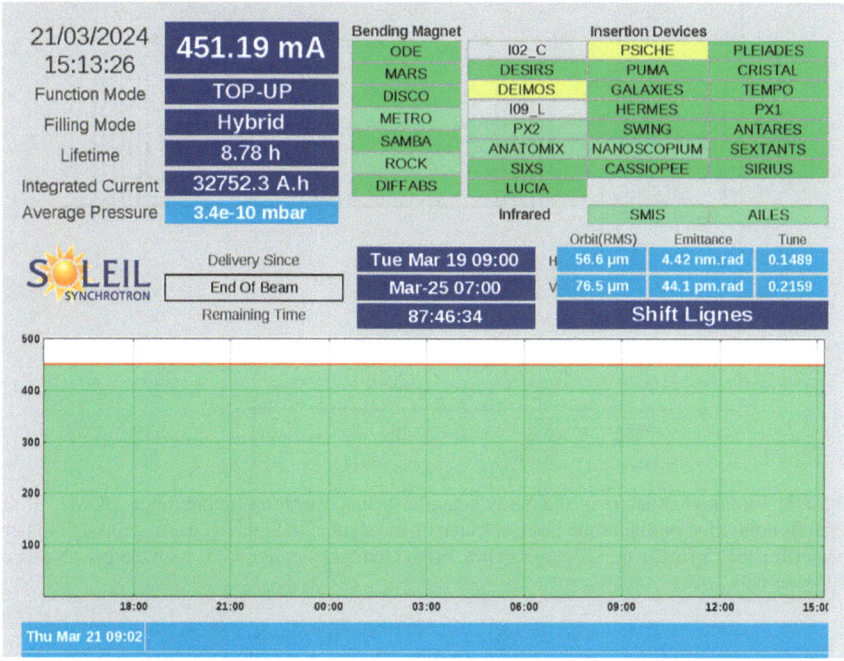

Fig. 1.2 The Soleil synchrotron display. Reproduced with the permission of the Soleil synchrotron

an electric field parallel to the beam orbit alternating in polarity, usually sinusoidally, at high frequency (50–500 MHz).

The temporal structure of the X-ray beam reproduces exactly that of the electrons (or positrons) in the storage ring. The electrons travel in the ring in bunches and thus the radiation is emitted in pulses. The length of the bunch is determined partly by the frequency and amplitude of the accelerating field in the storage ring cavities and partly by the details of the magnet lattice. There are other processes which can cause the bunch to be longer than the natural length. One of these processes is the interaction on the bunch by fields induced by the bunch itself in the walls of the vacuum chamber. This has a characteristic signature in which the bunch length increases with the one-third power of the bunch current, above a threshold current. The natural bunch length is therefore only likely to be experienced at small beam currents. The bunch separation (typically 2–20 ns) is synchronised to the radiofrequency (r.f.) accelerating field frequency in the main storage ring (and booster if used) because the r.f. has to replenish the energy lost to SR. The circumference length of the machine can accommodate a certain maximum number of bunches called the harmonic number; when all these bunches are filled this is known as multibunch mode. It is also possible

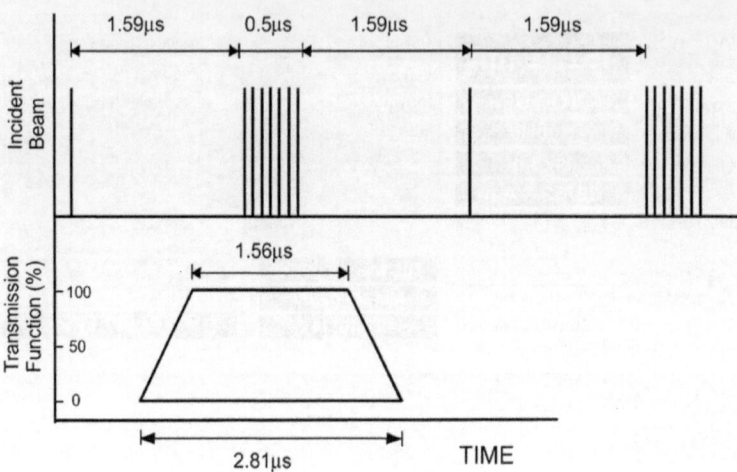

Fig. 1.3 Top: time structure of the X-ray beam in special operating timing mode 'SOM1' of the APS. Bottom: time profile of the beam selected by the shutter and thereby passing through to (and then diffracted by) a 0.1 mm crystal sample. From Gembicky et al (2005) with the permission of IUCr Journals

to operate the machine with only one bunch filled (i.e. single bunch mode). The time between light flashes is then determined by the orbital period, which is useful for time resolved experiments. Moffat (1987) suggested that if the number of photons provided by a single bunch of electrons was sufficient to measure a diffraction pattern, then the time resolution of the experiment would be the bunch length. The feasibility of this was demonstrated subsequently on CHESS using an undulator as the source (Szebenyi et al. 1988). On the accelerator side the machine physicists introduced other timing modes of the electron bunches while maintaining a high current operation. Low current mode is a limitation of single bunch machine operation. The hybrid filling mode consists of an isolated single bunch in an otherwise 'normal filling' multibunch mode. To some extent it is also possible to shorten the length of an electron bunch. Gembicky et al (2005) describe a shutter to select bunches in a special operating timing mode (SOM1) of the APS; see Fig. 1.3.

Typical values of the bunch width, separation and orbital period can be illustrated with respect to the Daresbury SRS and the ESRF in Grenoble. For the SRS the bunch width was 150 ps, the bunch separation 2 ns and the orbital period 321 ns. For the ESRF these values are respectively 65–140 ps, 3 ns and 2.84 μs. The orbital period of the ESRF is basically more than 8× larger than the SRS due to its much larger circumference.

Provided that various instabilities have thresholds at higher current values, the main limitation on the beam current that can be supported in a storage ring is set by the r.f. power available. The SR power emitted increases linearly with the beam current. The beam current in a storage ring decays with an exponential time constant but for many years now is replenished by a top up procedure of the machine ring current. Without top up, the lifetime of the beam is influenced by many factors. The dominant beam loss mechanism results from collisions of the electrons with residual gas molecules in the machine vacuum. Both inelastic and elastic scattering can take place off the nuclei and orbital electrons of the gas molecules. The beam lifetime is inversely proportional to the vacuum pressure that can be achieved. After the start-up of a new storage ring, or one which has been let up to atmosphere and then pumped down and baked, the lifetime will be poor. However, it will improve with operation. This is because the gas molecules adsorbed on the vacuum vessel surfaces are desorbed by the synchrotron radiation itself. Another process which can be of importance in limiting the lifetime is the scattering of one electron off another in the same bunch, known as the Touschek effect. The scattering rate depends on the electron energy and density in the bunch and so is important usually for low emittance or low energy machines, which have a high current in a short, small cross-section electron bunch. The advent of undulators requires the pole pieces of these insertion devices to be brought close together for short wavelength emission. There is a limit to how small the gap can be made because small apertures limit the lifetime primarily due to elastic Coulomb scattering of electrons off the residual gas molecules. At the early stages of operation of the ESRF, and indeed during its planning, the smallest undulator gap considered was 20 mm. Gaps as small as 5 mm are now routine.

For a collection of emitting charged particles (electrons or positrons) circulating repeatedly in a storage ring there is not just a single unique trajectory; there is a finite transverse distribution and angular spread of the beam controlled by the entire magnet lattice. The position and angular trajectory of an electron, and hence of the emitted photons, are correlated parameters. The phase space plot relates the position (x or y) of an electron and its angular trajectory (x' or y') with respect to the mean orbit; separate plots are needed for the vertical direction (y, y') perpendicular to the mean orbit plane and for the radial direction (x, x') in the orbit plane. The precise shape and orientation of these phase space ellipses changes for different points on the electron orbit but their area, the electron emittance, is invariant for a given machine lattice and electron energy. The emittance of the synchrotron radiation is calculated by convoluting the distribution of emitted photons with the electron emittance.

In terms of the principles of synchrotron radiation and the practicalities of producing it the radiation emanates with defined spectral characteristics from a finite source size and with a finite angular divergence.

The terms intensity, brightness and brilliance are used to describe the various aspects of SR and its use in experiments. The spectral flux is the number of photons emitted per unit time into a relative bandwidth into an angle element in the plane of the electron orbit and integrated in the vertical plane. The units of flux are usually number of photons/second/mrad horizontal/0.1% relative bandwidth. The intensity is used to define the flux per unit area of the wavefront some distance from the source. This is important when the sample cross section is smaller than the beam cross section. A diverging beam of 1 mrad horizontal × 0.2 mrad vertical, such as was the case with the SRS bending magnets and wigglers, will, in the absence of focussing, be 20 mm × 0.4 mm at 20 m from the source. For a 0.3 mm macromolecular crystal a high flux spread over such a large area of 8 mm^2 has a low intensity or put another way the sample intercepts only a small portion of the available flux. Focussing beamlines provide a higher intensity at the sample by collecting the available flux and focussing it at the sample; the small reflectivity losses of an X-ray optical element being more than compensated for by the large aperture of collection of the optics and the small focal spot. To produce a fine focal spot at the sample needs a small source size. Hence, another figure of merit for the source is source intensity. The source intensity is the flux divided by the source area. Thereby, at 3rd generation sources when dealing with undulator radiation, where you may have 20–50 μrad of diverging beam, unfocussed radiation even 70 m from the source offers a small size beam cross section at the crystal sample. Focussing is still used for the highest intensity needs.

The brightness of the source (Mills et al. (1994)) is then a useful figure of merit for which the default standard units are photons s^{-1} mm^{-2} $mrad^{-2}$ $(0.1\%$ bandwidth$)^{-1}$.

Regarding insertion devices (wigglers and undulators) and their radiation properties, these are different for what we have so far referred to with the simple circular orbital motion of electrons in bending magnets as sources of synchrotron radiation. Wigglers and undulators can be purposely designed to enhance specific characteristics of SR, namely extend the spectral range to shorter wavelengths, increase the available intensity, and provide an intense quasi-monochromatic beam. Less commonly used but still possible is to use an insertion device to provide a different polarization state of the beam. In insertion devices the electron beam oscillates as it experiences a periodic magnetic field. There is no net deflection or displacement and so such magnets can be inserted into straight sections between ordinary bending magnets. To extend the spectrum to shorter wavelengths the radius of curvature within a wiggler needs to be less than in a bending magnet for a given machine by using a higher magnetic field. Only a short straight section is needed for a three-pole wiggler, the simplest insertion device. To obtain light multiplication many magnetic poles are needed, and this is provided in either a multi-pole wiggler or an undulator, the former also usually acts as a wavelength shifter to shorter wavelengths. Long straight sections are needed for these devices. The classification of a periodic magnet insertion device as a wiggler or an undulator is based on whether the magnitude of the angular deflection of the electron beam is small enough to allow interference effects to be significant.

The addition of more magnet poles to a wiggler result in a multiplication of the flux generated. The flux multiplication factor is this number of poles. With an undulator extremely high fluxes of radiation are emitted at specific wavelengths. The interference conditions are met if the source emittance and the angular deflections of the electrons are very small. The undulator radiation is, as a result, tightly confined to the magnet axial direction and so can only serve one experiment at a time. Use of transparent monochromators do allow several stations to operate simultaneously with complementary radiation characteristics (Wakatsuki et al. 1998). The wavelengths of the emitted peaks depend on the machine energy as well as the magnetic field and pole period of the undulator. There is a gradual transition, as the magnetic field decreases, from multipole wigglers to undulators; at the same time the spectrum shifts to longer wavelengths and the electron deflection decreases. That is, it is possible for a single device to be operated in either mode.

An X-ray laser (XFEL) consists of highly collimated bunches of relativistic electrons from a long linear electron accelerator (LINAC) which then traverse a long magnetic undulator (for an overview see Patterson 2014). The major difference between synchrotron undulator radiation and an XFEL's undulator radiation is that the radiation field itself when traversing an 100 m XFEL undulator becomes sufficiently strong to influence the electron trajectory. The electrons whilst travelling very close to the speed of light are slightly slower. Whereas the electron bunch in a synchrotron consists of randomly distributed electrons, which thereby radiate incoherently, microbunching occurs in the XFEL resulting in coherent emission. This process is called self-amplifying spontaneous emission, which has its own acronym 'SASE'. The XFEL X-ray pulses are in the femtosecond range usually ~10 fs. There are complications to the pulse-to-pulse variation of their X-rays in spatial shape and width, which varies in a random way, compared with a synchrotron X-ray beam from a storage ring which is stable. Many experiments require narrow-bandwidth pulses with a well-defined wavelength. To help reduce these random effects at an XFEL it is possible to induce a 'seeding' of the beam using a seed pulse, e.g. from a UV laser or a high-harmonic generation source, coaxially with the electron beam in the undulator.

Table 1.1 lists the available SR and XFEL sources worldwide and their key beamline types that they provide as well as their weblinks.

Table 1.1 Synchrotron X-radiation and XFEL laboratories which have crystallography beamlines (given in alphabetical order of country). Further details on electron bioimaging facilities at several of the Synchrotron X-radiation laboratories are given in Table 20.1

Country	Facility	Stations	Webpage
Brazil	LNLS	MANACÁ (MAcromolecular micro and NAnoCrystAllography)	https://lnls.cnpem.br/facilities/manaca-en/
China	SSRF	BL17B1 High-throughput Protein Crystallography Beamline	http://e-ssrf.sari.ac.cn/beamlines/bl17b1/
China	HEPS	ID02 Microfocus protein crystallography	Microfocusing x-ray protein crystallography beamline–High Energy Photon Source (cas.cn)
China	HEPS	ID23 Time-resolved diffraction, imaging, and scattering	Structural dynamics beamline–High Energy Photon Source (cas.cn)
France	ESRF EBS	Many beamlines of diverse types; See especially Structural Biology (5 MX and 1 SAXS beamlines; cryoEM; an in crystallo-optical spectroscopy facility; a high-pressure MX cooling facility; a protein support laboratory	https://www.esrf.fr/UsersAndScience/Experiments/MX
France	ESRF EBS	Several Bending Magnet (BM) CRGs beamlines (Cooperative Research Groups)	https://www.esrf.fr/home/UsersAndScience/find-a-beamline.html
France	Soleil	Proxima I and II; MX, small molecule crystallography and powder diffraction	https://www.synchrotron-soleil.fr/en/beamlines/proxima-1 https://www.synchrotron-soleil.fr/en/beamlines/proxima-2a
Germany	EMBL	P13 and P14 MX; T-REXX pump-probe time-resolved crystallography	https://www.embl.org/groups/macromolecular-crystallography/

(continued)

Table 1.1 (continued)

Country	Facility	Stations	Webpage
Germany	BESSY2	MX	https://www.helmholtz-berlin.de/forschung/quellen/bessy/index_en.html and https://edoc.mdc-berlin.de/14996/
Germany	EuXFEL	Serial Crystallography	https://www.xfel.eu/science/scope/index_eng.html
India	INDUS	MX	Kumar et al. (2016). Protein crystallography beamline (PX-BL21) at Indus-2 synchrotron. J. Synchrotron Rad. 23, 629–634
Italy	Elettra	Biocrystallography; Powder diffraction; Surface diffraction; Time-resolved; High pressure	https://www.elettra.eu/lightsources/elettra/elettra-beamlines/diffraction-beamlines-at-elettra.html
Japan	PF	Structural biology	https://www.kek.jp/en/Facility/IMSS/PF/PFRing/PX/ https://lightsources.org/lightsources-of-the-world/asia-oceania/photon-factory/
Japan	SPring-8	Many beamlines of diverse types; including cryoEM	http://www.spring8.or.jp/en/about_us/whats_sp8/facilities/bl/list/ MX:- http://www.spring8.or.jp/wkg/BL41XU/instrument/lang-en/INS-0000000328/instrument_summary_view

(continued)

Table 1.1 (continued)

Country	Facility	Stations	Webpage
JAPAN	SACLA	Serial Crystallography	https://www.riken.jp/en/research/labs/rsc/adv_photon/sacla_bl/index.html
Russian Federation	Kurchatov Institute	Many beamlines of diverse types including MX and chemical crystallography	http://kcsni.nrcki.ru/pages/en/beamlines/index.shtml
Spain	ALBA	To MX beamlines; cryoEM	https://www.albasynchrotron.es/en/beamlines
Sweden	MAX-IV	Many beamlines of diverse types	https://www.maxiv.lu.se/beamlines-accelerators/beamlines/
Taiwan	Synchrotron Radiation Research Center	Many beamlines of diverse types	https://www.nsrrc.org.tw/english/index.aspx
UK	Diamond Light Source	Many beamlines of diverse types; See especially Structural Biology (5 MX beamlines and several cryoEMs at the eBic centre)	https://www.diamond.ac.uk/Home.html https://www.diamond.ac.uk/Instruments.html
USA	APS	Industrial Macromolecular Association (IMCA) of pharmaceutical companies, founded in 1990	https://www.aps.anl.gov/Beamlines/Directory https://www.imca.aps.anl.gov/About
USA	ALS	Many beamlines of diverse types	https://als.lbl.gov/beamlines/
USA	CAMD	Many beamlines of diverse types	https://www.lsu.edu/camd/beamlines/techniques.php
USA	CHESS	Many beamlines of diverse types; See especially MacCHESS Structural Biology	https://www.chess.cornell.edu/users/beamlines
USA	SSRL	Many beamlines of diverse types including 3 MX	https://www-ssrl.slac.stanford.edu/content/beam-lines/by-technique

(continued)

Table 1.1 (continued)

Country	Facility	Stations	Webpage
USA	LCLS II	Serial femtosecond crystallography and Pump probe	https://lcls.slac.stanford.edu/instruments
USA	NSLS II	Many beamlines of diverse types including 2 MX; and a cryoEM Centre	https://www.bnl.gov/nsls2/beamlines/ https://www.bnl.gov/nsls2/lifesciences/ https://www.bnl.gov/cryo-em/

References

Gembicky M, Oss D, Fuchs R, Coppens P (2005) A fast mechanical shutter for submicrosecond time-resolved synchrotron experiments. J Synchrotron Rad 12:665–669

Hwu Y, Margaritondo G (2021) Synchrotron radiation and X-ray free-electron lasers (X-FELs) explained to all users, active and potential. J Synchrotron Rad 28:1014–1029

Patterson BD (2014) Crystallography using an X-ray free-electron laser. Crystallogr Rev 20(4):242–294. https://doi.org/10.1080/0889311X.2014.939649

Wakatsuki S, Belrhali H, Mitchell EP, Burmeister WP, McSweeney SM, Kahn R, Bourgeois D, Yao M, Tomizaki T, Theveneau P (1998) ID14 'Quadriga', a beamline for protein crystallography at the ESRF. J Synchrotron Rad 5:215–221

Chapter 2
A Beginner's Guide to Neutron Reactor and Spallation Sources

Neutrons used in crystallography research are generated in two types of sources: nuclear reactors and spallation neutron sources. While reactors produce neutrons continuously, spallation sources produce pulses of neutrons due to the pulsed nature of the particle accelerators used. The generated neutrons have energies of mega electron volt (MeV) and thus have wavelengths that would be extremely short for our diffraction purpose. The emitted neutrons are therefore passed through materials known as "moderators" to reduce the neutron energies to the milli electron volt (meV) range. According to the type of moderator the neutron beams are called "cold" having wavelengths of ~ 5 Å or "thermal" or "hot" neutrons having wavelengths of ~ 1 Å. Table 16.1 provides a tabulation of where the neutron sources are around the world that have crystallography, diffraction and scattering instruments. An extensive description of the production of neutrons and the technology to deliver neutrons to the instruments is given by Niimura and Podjarny (2011). A wide range of references and books, and a history of the production and uses of neutrons in structure of matter research, is available at https://neutronsources.org/about/history/references/.

Neutron detectors encounter the challenge that their absorption of neutrons is weak unless there is a careful selection of a particular isotope such as helium-3 (He-3), lithium-6 (Li-6), boron-10 (B-10) or gadolinium (Gd). These are called neutron converters. An example is the neutron macromolecular crystallography instrument 'NMX' at the European Spallation Source (ESS) (Table 16.1) where Gd of a particular isotope increases the absorption by a factor of 3, from ~ 15% to ~ 45%; an up-to-date description of these aspects is given by Pfeiffer (2024).

In crystallography neutrons and X-rays are complementary probes; one starts with the X-ray crystal structure and then moves on to complete the hydrogenation/protonation details of the structure with neutrons (e.g. see Blakeley et al 2004 for macromolecular crystallography). The complementarity of electrons as probe (microED or cryoEM) with neutrons is as yet unexplored.

© The Author(s), under exclusive license to Springer Nature Switzerland AG 2025 13
J. R. Helliwell, *Certifying Central Facility Beamlines for Biological and Chemical Crystallography and Allied Methods*,
SpringerBriefs in Crystallography, https://doi.org/10.1007/978-3-031-80181-5_2

References

Blakeley MP, Cianci M, Helliwell JR, Rizkallah, PJ (2004) Synchrotron and neutron techniques in biological crystallography. Chem Soc Rev 548–557

Niimura N, Podjarny A (2011) Neutron protein crystallography. IUCr Book Series, Published by Oxford University Press, Oxford, UK

Pfeiffer D and the neutron time-of-flight collaboration (2024). https://cds.cern.ch/record/2886155/files/INTC-I-271.pdf

Chapter 3
A Beginner's Guide to Detectors in Use at X-Ray Beamlines and Their Calibration

At the beamline it is generally assumed that the electronic area detector's calibration has been undertaken by the supplying manufacturer and/or beamline scientist. These calibrations include correcting for its non-uniformity of intensity response across its area and its non-linearity of intensity measurements as well as spatial distortions. Alkire et al. (2016) at the USA APS in Chicago on the Structural Biology sector investigated these calibrations for their CCD detector, a device based on conversion of X-ray photons to visible photons by a phosphor layer. They found that there were imperfections in the non-uniformity calibration. In another study this type of error was found and corrected for using an ad hoc post-measurement procedure by Sanders et al. (2001) for diffraction data measured on the IMCA sector also at APS. The discovery of an anomaly in the atomic B factors was seen by Sanders et al. (2001) but successfully corrected by scaling the Fobs to Fcalcs using SFALL (CCP4 1994) from an intermediate protein model. The Protein Data Bank finalised deposition has accession code 1I3H.

Working with the manufacturers of detectors directly, before or after installation at a beamline, I have found productive; see for example the experiences of calibration and recalibration of the Nicolet Xentronics (Derewenda and Helliwell (1989)).

A versatile method for uniformity calibration of an electronic area detector, applicable to beamlines providing a wide range of photon energies from 5 to 50 keV, was developed at ESRF by Moy et al. (1996)) and expanded on in Hammersley et al. (1997). They used doped lithium borate glasses to obtain a smooth, almost isotropic, illumination of a two-dimensional detector for flat-field correction.

A user may reasonably ask the beamline staff *"how does your beamline's detector compare with the ideal performance detector especially with weak signals?"*. This can be illustrated with Fig. 3.1. For a description of properties of detectors and their characterisation see the book (Helliwell 1992) Sect. 5.4. A further description of the properties of detectors for X-ray diffraction studies, as well as the detailed comparison of the performances of different detector types is provided by Gruner

J. R. Helliwell, *Certifying Central Facility Beamlines for Biological and Chemical Crystallography and Allied Methods*, SpringerBriefs in Crystallography, https://doi.org/10.1007/978-3-031-80181-5_3

The detector challenge
The ideal detector
Photon counting detector (top) vs Image plate (bottom)
with weak signals: 10 & 0.3 secs exposure.
Extend the diffraction resolution via a better detector!

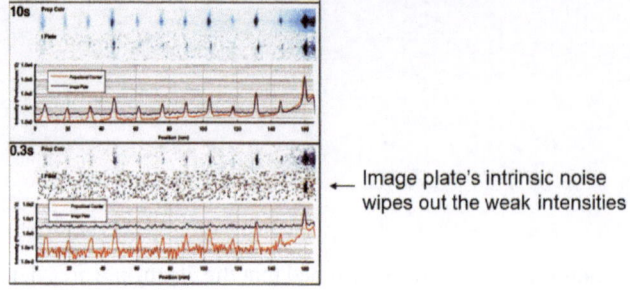

← Image plate's intrinsic noise
wipes out the weak intensities

Rob Lewis, Daresbury Laboratory J. Synchrotron Rad. (1994). 1, 43-53

Fig. 3.1 This work of Dr Rob Lewis at the Synchrotron Radiation Source at Daresbury Laboratory, UK, showed what one is striving for in terms of the ideal photon counting detector compared with the best integrating (analogue) detector of the time the image plate. Reproduced with the permission of IUCr Journals from Lewis (1994) *Multiwire Gas Proportional Counters: Decrepit Antiques or ClassicPerformers?* J Synchrotron Radiation 1, 43–53

et al. (2012). A predominant supplier of detectors for SR beamlines is the company Dectris with their range of pixel devices described at https://www.dectris.com/en/.

References

Alkire RW, Rotella FJ, Duke NEC, Otwinowski Z, Borek D (2016) Taking a look at the calibration of a CCD detector with a fiber-optic taper. J Appl Cryst 49:415–425

CCP4 (1994) The CCP4 suite: programs for protein crystallography. Acta Crystallog. Sect. D 50:760–763

Derewenda Z, Helliwell JR (1989) Calibration tests and use of a Nicolet/Xentronics imaging proportional chamber mounted on a conventional source for protein crystallography. J Appl Cryst 22:123–137

Gruner SM, Eikenberry EF, Tate MW (2012) Chapter 7.1. Comparison of X-ray detectors. Int Tables Crystallogr F(ch. 7.1):177–182 |1|2|. https://doi.org/10.1107/97809553602060000820

Hammersley AP, Brown K, Burmeister W, Claustre L, Gonzalez A, McSweeney S, Mitchell E, Moy J-P, Svensson SO, Thompson AW (1997) Calibration and application of an X-ray image intensifier/charge-coupled device detector for monochromatic macromolecular crystallography. J Synchrotron Rad 4:67–77

Helliwell JR (1992) Macromolecular crystallography with synchrotron radiation. Cambridge University Press Section 5.4

Lewis R (1994) Multiwire gas proportional counters: decrepit antiques or classic performers? J Synchrotron Radiation 1:43–53

Moy JP, Hammersley AP, Svensson SO, Thompson A, Brown K, Claustre L, Gonzalez A, McSweeney S (1996) A novel technique for accurate intensity calibration of area X-ray detectors at almost arbitrary energy. J Synchrotron Rad 3:1–5

Sanders DAR, Moothoo DN, Raftery J, Howard AJ, Helliwell JR, Naismith JH (2001) The 1.2 Å resolution structure of the Con A-dimannose complex. J Mol Biol 310:875–884

Chapter 4
Introduction to X-Ray Beamlines for Crystallography

Beam line front ends are vital for protecting the machine vacuum. The lifetime of the beam in the ring depends on the vacuum level. Hence, the beam lines emanating from the ring are separated from the main ring vacuum chamber. For an X-ray beam line this separation is typically by using a beryllium window at the front end of the beam line. For a soft X-ray or vacuum ultra-violet beamline such a window would completely absorb the radiation of that wavelength so instead a fast valve is placed in the beam line to close quickly in the event of a vacuum failure in the beam line. Water-cooled absorbers and masks are also placed in the beam line front end. There are fixed and movable masks. A fixed mask is located next to the bending magnet vacuum chamber; it constrains the opening fan of bending magnet radiation. The movable masks are the personal safety shutters used to close off the beam when access to an experimental hutch is required. Insertion devices present additional problems associated with the extremely high heat loads. These can melt the metal components in valves in the event of a malfunction. In normal operation the front-end masks are constructed with absorbing surfaces at a large angle with respect to the beam so that the power is absorbed across a large surface area. In this way only the radiation from the central undulator cone is used by the beamline, giving a lower heat loading on optical components.

The monitoring and stability of the beam are essential for which photon beam position monitors ensure that after an injection the electron beam position is adjusted to allow the synchrotron radiation to strike the beam line optical components in a constant way. The wavelength output from a double crystal monochromator is especially sensitive to the vertical beam position. Also, the quality of the focus from a toroid mirror is especially sensitive to the horizontal beam position. It used to be necessary to re-calibrate the wavelength and the focussing of a beamline optical system after each beam injection. Slow position drifts are associated with temperature variations of the magnet lattice. Irregular jumps are usually due to steering magnet power supply instabilities; these jumps used to be as large as 50 μm on the second-generation SR sources even though dedicated to SR production. Obviously

J. R. Helliwell, *Certifying Central Facility Beamlines for Biological and Chemical Crystallography and Allied Methods*, SpringerBriefs in Crystallography, https://doi.org/10.1007/978-3-031-80181-5_4

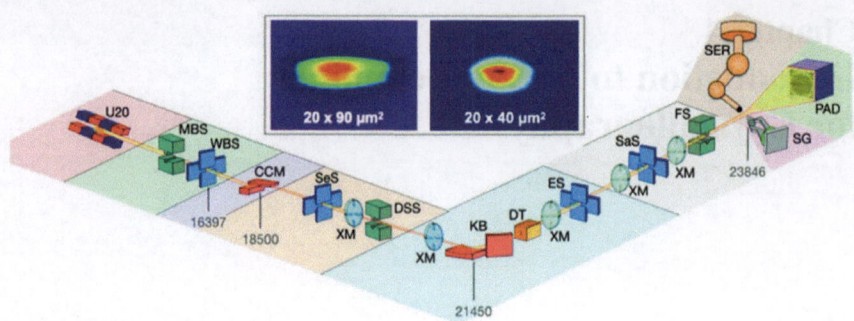

Fig. 4.1 General layout of Soleil's PROXIMA-1 beamline for macromolecular crystallography (from Chavas et al (2021)). Relative distances (not to scale) are from the light source down to the sample. U20, *in vacuum* undulator; MBS, main beam shutter; WBS, white-beam slits; CCM, channel-cut monochromator; SeS, secondary slits; DSS, downstream safety shutter; KB, KB mirrors; DT, diagnostic tools; ES, experimental slits; SaS, sample slits; FS, fast shutter; SG, SmarGon; PAD, pixel-array detector; SER, sample exchange robot; XM, X-ray beam positioning monitor. Elements are colour-located by functional blocks. The top inlet represents the direct beam at the sample position scaled and imaged on a motorized imager, with (right) and without (left) a diaphragm. Reproduced from Chavas et al (2021) with the permission of IUCr Journals and the authors

for a focal spot of 250 μm, typical at the SRS, a 50 μm source position jump results in a corresponding shift in the focal spot position. Hence, the beam flux transmitted through a pinhole collimator would jump dramatically and, if not corrected for, introduce errors into the experimental data. The position and angle stabilities obtained in the 1990s were much improved and became better than 10 μm and a few microradians respectively over short and long term. Today, at the time of writing this Brief Book, these stability performances have further improved considerably. Beams at the crystal are now usually smaller than 50 μm. Another feature of preserving beam stability at the sample position is a constant heat loading onto the beamline optics, which is achieved by having the X-ray optics warm continuously, as well as storage ring top up injection which keeps the circulating synchrotron beam current stable to around 1%. The biggest impact on beam stability nowadays is beam angle through the 'pinhole' defined by slits, mirror acceptance and different apertures. For an 'EBS', (Extremely Bright Source) the aim is to keep source stability to be under i.e. better than ± 1–2 μm and ± 1 μrad.

There has been a considerable evolution of beamlines for crystallography with various types consistently provided at the facilities. I give specialist chapter descriptions of them below. For an example see Fig. 4.1.

Reference

Chavas LMG, Gourhant P, Guimaraes BG, Isabet T, Legrand P, Lener R, Montaville P, Sirigu S, Thompson A (2021) PROXIMA-1 beamline for macromolecular crystallography measurements at synchrotron SOLEIL. J Synchrotron Rad 28:970–976

Chapter 5
Guidance in Preparing the Experiment at Your Home Facility Including Sample Preparation

First and foremost, of course, is crystallisation for the crystallographic study (see e.g. Chayen et al. 2010) and single particle grids for cryoelectron microscopy (cryoEM) work (see e.g. Weissenberger et al 2021). Secondly, an evaluation of whether your crystal diffracts, and whether it cools in an acceptable manner for cryo work should be done. Likewise, for cryoEM it can be the case that the access to a 'high end microscope' at a central facility requires demonstration of the quality of the grids. The ideals sought are that the thickness be not to thick, not too thin and the particles are intact. As an example of the access requirements for approval of the use of a high end microscope see the ESRF web pages https://www.esrf.fr/cryoEM-SOS [esrf.fr] and Fig. 20.1.

Special requirements exist for neutron crystallography that a protein crystal will need to be larger than around $0.5 \times 0.5 \times 0.5$ mm^3. This may be explained at the neutron facility instrument website as a crystal sample volume value such as: > 0.1 mm^3. Fully deuterated protein will allow use of a smaller crystal than a ^{1}H/^{2}H partially exchanged protein as the incoherent scattering from deuterium is minimal compared with that of protium and the large number of deuteriums in a protein add considerably to the scattering strength of the crystal. Neutron facilities, and the instrument local contact, should be consulted before a beamtime proposal is made as they will have in depth practical experience of whether the protein crystal unit cell volume is out of scope of their instrument i.e. too large. Some of the neutron crystallography instruments worldwide can accept unit cells up to about $300 \times 300 \times 300$ Å^3.

Special sample requirements exist at the XFELS where a fluid or gel-based stream of micron or sub-micron sized crystals are needed e.g. for suitable photoactivatable reactions in crystals. Fixed target approaches may supersede this approach and have already had strong scientific results.

Micro electron diffraction (MicroED) has a need for a crystal to be within a range of approximately 200–400 nm thickness due to the very strong scattering interaction of electrons with matter. This size range is close to the wavelength of visible light

J. R. Helliwell, *Certifying Central Facility Beamlines for Biological and Chemical Crystallography and Allied Methods*, SpringerBriefs in Crystallography, https://doi.org/10.1007/978-3-031-80181-5_5

and therefore make them difficult to see with light microscopes. When placed on the EM grid, if they have plate like crystal morphology, such crystals will sit on the grid in a single orientation and simply tilting the grid will not be sufficient to collect a complete electron diffraction data set. Some facilities have specialised devices to deal with this challenge (Heidler et al 2019). When the crystal size is not small enough then focussed ion beam milling can be used to reduce their thickness. The extent of damage to the crystal using different focussed ion beams has been evaluated by Parkhurst et al (2023) and found to be not limiting.

Sample preparation methods for cryoEM are more mature than the XFEL sample delivery case but are still under active development; for an example see Joppe et al. (2020). Cryo electron tomography for in-cell imaging requires the sample to be 'cut' to a thin enough thickness. Streams of liquid gallium metal ions in a beam can be used to shave the cell to a suitable thickness.

References

Chayen NE, Helliwell JR, Snell EH (2010) Macromolecular crystallization and crystal perfection (International Union of Crystallography Monographs on Crystallography). Published by Oxford University Press. ISBN-9780199213252. https://doi.org/10.1093/acprof:oso/9780199213252.001.0001

Heidler J, Pantelic R, Wennmacher JTC, Zaubitzer C, Fecteau-Lefebvre A, Goldie KN, Muller E, Holstein JJ, van Genderen E, De Carlo S, Gruene T (2019) Design guidelines for an electron diffractometer for structural chemistry and structural biology. Acta Cryst D75:458–466

Joppe M, D'Imprima E, Salustros N, Paithankar KS, Vonck J, Grininger M, Kuhlbrandt W (2020) The resolution revolution in cryoEM requires high-quality sample preparation: a rapid pipeline to a high-resolution map of yeast fatty acid synthase. IUCrJ 7:220–227

Parkhurst JM, Crawshaw AD, Siebert CA, Dumoux M, Owen CD, Nunes P, Waterman D, Glen T, Stuart DI, Naismith JH, Evans G (2023) Investigation of the milling characteristics of different focused-ion-beam sources assessed by three-dimensional electron diffraction from crystal lamellae. IUCrJ 10:270–287

Weissenberger G, Henderikx RJM, Peters PJ (2021) Understanding the invisible hands of sample preparation for cryo-EM. Nat Methods 18:463–471. https://doi.org/10.1038/s41592-021-01130-6

Chapter 6
Beamline Scientist Perspective

Aragao and Cowieson (2022) describe the checks which the beamline scientists do, including daily, for their beamlines. They nicely describe quality control systems which give a maintenance record of checks with the following advantages:

- optimizing beamtime by guaranteeing that certain tests are done in regular intervals;
- plot beamline degradation or improvements particularly when new software or hardware is implemented;
- guarantee that beamline performance is not dependent of which synchrotron staff do the checks because they are all done the same way.

For beam characterisation, at a minimum the beamline scientist will do knife edge scans for beam size and will have calibrated diodes for flux measurements (see e.g. Owen et al 2009). The facility's optics group will have fast cameras for visualising the beam as well and the facility's machine control room will have X-ray beam position monitors on many of the beamlines that can be analysed.

References

Aragao D, Cowieson N (2022) Beamline setup and calibration quality control for synchrotron MX beamlines. Acta Cryst A 78:e789
Owen RL, Holton JM, Schulze-Briese C, Garman EF (2009) Determination of X-ray flux using silicon pin diodes. J Synchrotron Rad 16:143–151

SpringerBriefs in Crystallography, https://doi.org/10.1007/978-3-031-80181-5_6

Chapter 7
Beamline User Perspective

The user predominantly is focussed on their sample. They will be as certain as they can be from their several biophysical and biochemical characterization methods that they know their sample. In two ways this has been called into question. Firstly, Garman and Grime (2005, and references therein) introduced PIXE (Proton Induced X-ray Emission) to be sure of what metal type is in a given biological macromolecule. In collaboration with the Hauptmann Woodward Institute's USA National Crystallisation Facility in Buffalo, Grime et al. (2020) conducted a post publication/PDB deposition analysis of crystal samples still held at that Facility. They found from bond lengths' analyses, along with their new protein model's difference Fourier electron density maps being cleaner, that a fraction of some 20% had PDB models with the wrong metal type and which they had corrected based on their PIXE analyses. Some MX synchrotron facilities have introduced routine fluorescence X-ray scans at their beamlines to provide a correctly identified metal information to the user (Leonard et al. 2009).

A beamline user over the decades has seen an evolution of detector hardware and software. On the fly analyses of the raw diffraction images is now routine. Furthermore, the remote access has provided a "*no need to travel to the facility*", a major and largely welcome change of user-lifestyle. This has also increased efficiency with more routine turnaround of users and the ability to redistribute unused beamtime if samples do not diffract as planned. An early initiative of remote access technology and use is described here by Smith et al. (2010) and is available at all the synchrotron facilities. Many years later, this proved vital for keeping the quantity of Covid-19 macromolecular crystallography data collections going at pace. Bowler et al. (2016) have pioneered a fully automatic MX facility at the ESRF.

In terms of ease of use the customised systems such as MxCUBE (Gabadinho et al. 2010) allow users to interact with beamline hardware components easily and simply. It also provides automated routines for common tasks in the operation of a synchrotron beamline dedicated to experiments in MX. MxCUBE is widely available e.g. used at ESRF, Elettra, Petra III and BESSY2.

© The Author(s), under exclusive license to Springer Nature Switzerland AG 2025   27
J. R. Helliwell, *Certifying Central Facility Beamlines for Biological and Chemical Crystallography and Allied Methods*,
SpringerBriefs in Crystallography, https://doi.org/10.1007/978-3-031-80181-5_7

After a data collection run data processing pipelines present the results of the most commonly available softwares with a single set of raw diffraction images. With well behaved crystals they naturally produce very similar analysis outcomes. With more challenging crystal samples showing up such as split spots, order disorder effects, or diffraction anisotropy the softwares available tend to show a variance of outcomes. See Chap. 26.

References

Bowler MW, Svensson O, Nurizzo D (2016) Fully automatic macromolecular crystallography: the impact of MASSIF-1 on the optimum acquisition and quality of data. Crystallogr Rev 22(4):233–249. https://doi.org/10.1080/0889311X.2016.1155050

Gabadinho J, Beteva A, Guijarro M, Rey-Bakaikoa V, Spruce D, Bowler MW, Brockhauser S, Flot D, Gordon EJ, Hall DR, Lavault B, McCarthy AA, McCarthy J, Mitchell E, Monaco S, Mueller-Dieckmann C, Nurizzo D, Ravelli RBG, Thibault X, Walsh MA, Leonard GA, McSweeney SM (2010) MxCuBE: a synchrotron beamline control environment customized for macromolecular crystallography experiments. J Synchrotron Rad 17:700–707

Garman EF, Grime GW (2005) Elemental analysis of proteins by microPIXE. Prog Biophys Mol Biol 89:173–205

Grime GW, Zeldin OB, Snell ME, Lowe ED, Hunt JF, Montelione GT, Tong L, Snell EH, Garman EF (2020) High-throughput PIXE as an essential quantitative assay for accurate metalloprotein structural analysis; development and application. J Am Chem Soc 142(1):185–197

Leonard GA, Sole VA, Beteva A, Gabadinho J, Guijarro M, McCarthy J, Marrocchelli D, Nurizzo D, McSweeney S, Mueller-Dieckmann C (2009) Online collection and analysis of X-ray fluorescence spectra on the macromolecular crystallography beamlines of the ESRF. J Appl Cryst 42:333–335

Smith CA, Card GL, Cohen AE, Doukov TI, Eriksson T, Gonzalez AM, McPhillips SE, Dunten PW, Mathews II, Song J, Soltis SM (2010) Remote access to crystallography beamlines at SSRL: novel tools for training, education, and collaboration. J Appl Cryst 43:1261–1270

Chapter 8
Fixed Wavelength High Intensity Beamline

Fixed wavelength beamlines have been established as they are somewhat easier and less costly to build and operate, as well as provide a higher intensity, than tuneable beamlines. They have increasingly come into fashion as protein structure determination using molecular replacement programs in macromolecular crystallography have been easier to use and, especially, the number of unique protein folds in the PDB have grown substantially. The arrival of AlphaFOLD and RosettaFold are much more recent developments (see Jumper et al. (2021) and Baek et al. (2021) respectively). Their use as a starting molecular replacement model was quickly identified (Helliwell (2020, 2021) and Gildea et al. (2022)) and has further increased the usage of fixed wavelength beamlines.

For effective raw diffraction images data processing the X-ray wavelength must be set and known to a good precision, typically 1×10^{-3}. This is not as precise as required for setting the wavelength to optimise anomalous dispersion (f′ and f″ values), typically 1×10^{-4}. See next Chapter. Some fixed wavelength stations, and others that are variable, have been exploring larger band passes with gradient index monochromators to increase flux (see Chap. 10 below).

The other critical factors for good raw diffraction images processing are that the crystal to detector angle, the detector placement details, and the direct beam position in the diffraction image must be precisely known. The latter was identified early in the development of synchrotron crystallography beamlines as an important certification of beamline performance for successful large unit cells work (Helliwell and Thompson (1983)). The X-ray wavelength is readily measured and set by using a pure metal foil (see e.g. Helliwell et al. (1982)). A further discussion of careful beamline settings involving fixed wavelength SAXS beamlines is described by Liu and Li (2013). The use of multiple detector positions separated by a known distance assist the precise estimation of wavelength using a standard sample and is described by Horn et al. (2019), who used a standard powder. The most precise calibration of an X-ray wavelength, to 5 decimal places, uses four beam diffraction (Huang

© The Author(s), under exclusive license to Springer Nature Switzerland AG 2025
J. R. Helliwell, *Certifying Central Facility Beamlines for Biological and Chemical Crystallography and Allied Methods*,
SpringerBriefs in Crystallography, https://doi.org/10.1007/978-3-031-80181-5_8

et al. 2022). Standard reference silicon powder is available from the USA National Institute of Standards and Technology (NIST; https://www.nist.gov/srm).

In the modelling of X-ray radiation damage of a protein crystal it became apparent that the incident beam profile would better be rectilinear and not Gaussian in shape (Zeldin et al. (2013)). The beamline can be configured that way (Spiga 2018).

The XFELs require their own, as well as features common to synchrotron sources, photon beam characterisation notably because of pulse-to-pulse variations in beam profile shape and intensity at the crystal sample. Thus, photon diagnostics devices such as developed for SwissFEL (Juranic et al. 2018, 2023) have been built to anticipate the demands of both users and machine operators, and provide access to the nondestructive characterization of XFEL photon flux, position, spectrum, pulse length and arrival time.

References

Baek M et al (2021) Accurate prediction of protein structures and interactions using a three-track neural network. Science 373(6557):871. https://doi.org/10.1126/science.abj8754

Gildea RJ, Orr CM, Paterson NG, Hall DR (2022) Embedding AI in the protein crystallography workflow. Synchrotron Rad News 35(4):51–54. https://doi.org/10.1080/08940886.2022.2114723

Helliwell JR, Greenhough TJ, Carr PD, Rule SA, Moore PR, Thompson AW, Worgan JS (1982) Central data collection facility for protein crystallography, small angle diffraction and scattering at the Daresbury SRS. J Phys E 15:1363–1372

Helliwell JR, Thompson AW (1983) A remote-control direct-beam exposure device for the Arndt-Wonacott rotation camera. J Appl Cryst 16:579

Helliwell JR (2020) DeepMind and CASP14 IUCr newsletter (ISSN 1067-0696), vol 28, no 4, page 6. https://www.iucr.org/news/newsletter/volume-28/number-4/deepmind-and-casp14

Helliwell JR (2021) IUCr newsletter. https://www.iucr.org/news/newsletter/volume-29/number-2/reaction-to-announcement-of-alphafold-database

Horn C, Ginell KM, Von Dreele RB, Yakovenko AA, Toby BH (2019) Improved calibration of area detectors using multiple placements. J Synchrotron Rad 26:1924–1928

Huang X, Shi X, Assoufid L (2022) X-ray beam monitoring and wavelength calibration using four-beam diffraction. J Synchrotron Rad 29:159–166

Jumper J et al (2021) Highly accurate protein structure prediction with AlphaFold. Nature. https://doi.org/10.1038/s41586-021-03819-2

Juranic P, Rehanek J, Arrell CA, Pradervand C, Ischebeck R, Erny C, Heimgartner P, Gorgisyan I, Thominet V, Tiedtke K, Sorokin A, Follath R, Makita M, Seniutinas G, David C, Milne CJ, Lemke H, Radovic M, Hauri CP, Patthey L (2018) SwissFEL Aramis beamline photon diagnostics. J Synchrotron Rad 25:1238–1248

Juranic P, Alarcon A, Ischebeck R (2023) Online absolute calibration of fast FEL pulse energy measurements. J Synchrotron Rad 30:500–504

Liu J, Li Z (2013) Using a standard sample to estimate the X-ray wavelength of the 1W2A SAXS beamline at BSRF. J Synchrotron Rad 20:729–733

Spiga D (2018) X-ray beam-shaping via deformable mirrors: surface profile and point spread function computation for Gaussian beams using physical optics. J Synchrotron Rad 25:123–130

Zeldin OB, Gerstel M, Garman EF (2013) RADDOSE-3D: time- and space-resolved modelling of dose in macromolecular crystallography. J Appl Cryst 46:1225–1230

Chapter 9
Tuneable Wavelength Beamline

Abstract These beamlines seek to exploit the changes in the resonant X-ray scattering of specific atom or atoms in a crystal structure.

These beamlines seek to exploit the changes in the resonant X-ray scattering of specific atom or atoms in a crystal structure. There are equivalent effects with neutrons but not widely developed as they are smooth changing with neutron wavelength compared with what can be achieved at an X-ray absorption edge. An excellent source book on these opportunities and the experiments of the time, pre-synchrotron radiation, is the 1975 IUCr book (Ramaseshan and Abrahams (1975)). A good discussion and illustration of the importance of wavelength stability for optimising anomalous dispersion effects, using the EMBL Hamburg beamline X31, is described by Evans and Wilson (1999) who also provide a good overview of the field of solving the phase problem in MX by such methods up to 1999. A very broad survey across MX and chemical crystallography is provided in the extensive review of Cianci et al (2005). A very recent extensive literature review is given by Hendrickson (2023) to accompany his IUCr 2023 Ewald Prize Award.

In seeking the most precise anomalous differences at a chosen wavelength a measuring strategy that has developed is to measure Bijvoet differences on the same diffraction image. This minimises the time difference for each reflection intensity measurement passing through the Ewald sphere; Nieh and Helliwell (1995) quantify this. An early use of element specific anomalous differences, and careful alignment of the protein crystal, was at SRS 7.2 to establish the identity of a binuclear metal site comprising Mn and Ca in pea lectin (Einspahr et al. (1985), see their Fig. 1). The method employed was labour and beamtime intensive involving a single crystal rotation axis, multiple setting diffraction images and adjusting the crystal sample orientation on a goniometer head. A more effective approach is to have multiple axes crystal sample goniostat and then optimise the procedure to adjust the crystal setting angles; see e.g. White et al (2018) who very reasonably state that the "*procedure for all axes of a kappa goniometer can be carried out on the beamline in a matter of minutes, as a minimum of only four diffraction images from a well diffracting test sample need to be collected.*"

© The Author(s), under exclusive license to Springer Nature Switzerland AG 2025   31
J. R. Helliwell, *Certifying Central Facility Beamlines for Biological and Chemical Crystallography and Allied Methods*,
SpringerBriefs in Crystallography, https://doi.org/10.1007/978-3-031-80181-5_9

The range of X-ray wavelengths used has steadily expanded. High photon energies were proposed in the ESRF Foundation Phase Report in 1987 and implemented in systematic studies by Jakoncic et al (2006), and references therein. Longer wavelengths have become increasingly popular as shown in Cianci et al (2001). A spectacular piece of instrumentation is the Diamond Light Source's Long Wavelength beamline, including a Dectris pixel detector placed in vacuum to access even the phosphorus K edge (https://www.diamond.ac.uk/Instruments/Mx/I23.html); this is a unique facility worldwide. The most popular element is selenium in the Se-methionine initiative of Wayne Hendrickson (see the frequency of elements used in phasing in Table 1 of Hendrickson (1999)).

References

Cianci M, Rizkallah PJ, Olczak A, Raftery J, Chayen NE, Zagalsky PF, Helliwell JR (2001) Structure of lobster apocrustacyanin A_1 using softer X-rays. Acta Cryst D 57:1219–1229

Cianci M, Helliwell JR, Helliwell M, Kaucic V, Logar NZ, Mali G, Tusar NN (2005) Anomalous scattering in structural chemistry and biology. Crystallogr Rev 11:245–335

Evans G, Wilson KS (1999) A MAD experiment performed at the white line of the iridium LIII absorption edge in lysozyme. Acta Cryst D55:67–76

Hendrickson WA (1999) Maturation of MAD phasing for the determination of macromolecular structures. J Synchrotron Rad 6:845–851

Hendrickson WA (2023) Facing the Phase Problem. IUCrJ 10:521–543

Jakoncic J, Di Michiel M, Zhong Z, Honkimaki V, Jouanneau Y, Stojanoff V (2006) Anomalous diffraction at ultra-high energy for protein crystallography. J Appl Cryst 39:831–841

Nieh YP, Helliwell JR (1995) Time–differences between Friedel reflections: accuracy of crystal setting and requirements on beam stability. J Synchrotron Rad 2:79–82

Ramaseshan S, Abrahams SC (eds) (1975) Anomalous Scattering. Published by Munksgaard, Copenhagen

White KI, Bugris V, McCarthy AA, Ravelli RBG, Csanko K, Cassetta A, Brockhauser S (2018) Calibration of rotation axes for multi-axis goniometers in macromolecular crystallography. J Appl Cryst 51:1421–1427

Chapter 10
Laue Diffraction Beamline

Abstract There are quite a variety of configurations for this type of beamline. Principally these are categorised on whether they provide a wide wavelength bandpass (e.g. 0.5–2 Å) or a narrow wavelength bandpass (e.g. 0.9–1.1 Å). The latter can also be called 'pink beams' or in neutron Laue Diffraction beamlines, 'quasi-Laue'.

There are quite a variety of configurations for this type of beamline. Principally these are categorised on whether they provide a wide wavelength bandpass (e.g. 0.5–2 Å) or a narrow wavelength bandpass (e.g. 0.9–1.1 Å). The latter can also be called 'pink beams' or in neutron Laue Diffraction beamlines, 'quasi-Laue'. A way to produce a pink X-ray beam bandpass is by use of a multilayer (see e.g. Kazimirov et al. (2006) and Fig. 10.1). A discussion of the choice of the bandpass of the X-ray source for Laue measurements is given by Šrajer et al. (2000).

In all cases of Laue diffraction, the unit cell parameters of a crystal sample need to be known i.e. calibrated first with a monochromatic beam. For time-resolved or perturbation studies this is not really a limitation because the 'static' or 'starting' crystal structure is known. For determining a new crystal structure, where the higher intensity of using a wide bandpass beam might be of interest, the question arises: is there a way of determining the cell parameters of the crystal sample from the Laue diffraction pattern alone? Medjoubi et al (2012) offer a way to do just that for a protein crystal when using a CdTe-XPAD detector. Send et al. (2009) also demonstrated this with a pnCCD type of electronic area detector, demonstrated for a crystal of γ-LiAlO$_2$.

An extensively calibrated suite of open-source Laue diffraction processing software is described by Hao et al (2021) and references therein.

J. R. Helliwell, *Certifying Central Facility Beamlines for Biological and Chemical Crystallography and Allied Methods*,
SpringerBriefs in Crystallography, https://doi.org/10.1007/978-3-031-80181-5_10

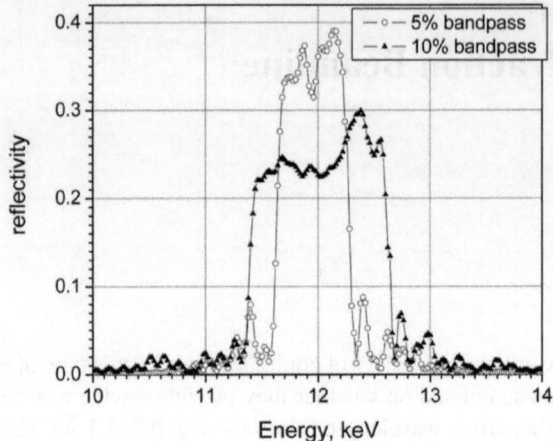

Fig. 10.1 Reflectivity curves from 5% (open circles) to 10% (triangles) broad-bandpass multilayers (MLs) measured by scanning the energy of an upstream Si (111) monochromator at fixed incident angle. These MLs were designed to have the energy bandpass centred at 12.0 keV at 1° incident angle of the synchrotron beam onto the ML. A typical ML comprises 500 layers of a material such as Al_2O_3/B_4C. Reproduced from Kazimirov et al (2006) with the permission of IUCr Journals

References

Hao Q, Harding MM, Helliwell JR, Szebenyi DM (2021) Weblinks for the Daresbury Laue software source code and information. J Synchrotron Rad 28:666

Kazimirov A, Smilgies D-M, Shen Q, Xiao X, Hao Q, Fontes E, Bilderback DH, Gruner SM, Platonov Y, Martynov V V (2006) Multilayer X-ray optics at CHESS. J Synchrotron Rad 13:204–210

Medjoubi K, Thompson A, Berar J-F, Clemens J-C, Delpierre P, Da Silva P, Dinkespiler B, Fourme R, Gourhant P, Guimaraes B, Hustache S, Idir M, Itie J-P, Legrand P, Menneglier C, Mercere P, Picca F, Samama J-P (2012) Energy resolution of the CdTe-XPAD detector: calibration and potential for Laue diffraction measurements on protein crystals. J Synchrotron Rad 19:323–331

Send S, von Kozierowski M, Panzner T, Gorfman S, Nurdan K, Walenta AH, Pietsch U, Leitenberger W, Hartmann R, Strader L (2009) Energy-dispersive Laue diffraction by means of a frame-store pnCCD. J Appl Cryst 42:1139–1146

Šrajer V, Crosson S, Schmidt M, Key J, Schotte F, Anderson S, Perman B, Ren Z, Teng T, Bourgeois D, Wulff M, Moffat K (2000) Extraction of accurate structure-factor amplitudes from Laue data: wavelength normalization with wiggler and undulator X-ray sources. J Synchrotron Rad 7:236–244

Chapter 11
Microfocus Beamlines

Abstract The impetus for this type of beamline arose out of planning the European Synchrotron Radiation Project (ESRP) which of course became the ESRF, the F being for Facility. In the Foundation Phase Report (ESRF 1987) the chapters on macromolecular crystallography (Helliwell (1987)) included one for microfocus. Microfocus beamlines provision has grown substantially at the 3rd generation SR sources.

The impetus for this type of beamline arose out of planning the European Synchrotron Radiation Project (ESRP) which of course became the ESRF, the F being for Facility. In the Foundation Phase Report (ESRF 1987) the chapters on macromolecular crystallography (Helliwell (1987)) included one for microfocus. The controversy for the time was the question put at ESRP Workshops: "*Would the protein crystal withstand the radiation and thermal blast of the ESRP beams?*". Evidence that they would withstand the beam was based on evaluation tests made in an SRS UK and SSRL USA collaboration. This used the SRS superconducting wiggler white beamline 9.6 which demonstrated good and sustained diffraction from (20 micron)^3 gramicidin crystals (Hedman et al (1985)). The wiggler white beam provided an integrated intensity at the crystal of around 10^{14} photons/sec/mm^2 and which matched the expected ESRF undulator monochromatic intensities. Microfocus beamlines provision has grown substantially at the 3rd generation SR sources (see e.g. the overviews over a historical timeline: Blakeley et al (2004), Flot et al (2006), Evans et al. (2011) and Hirata et al (2016)).

References

Blakeley MP, Cianci M, Helliwell JR, Rizkallah PJ (2004) Synchrotron and neutron techniques in biological crystallography. Chem Soc Reviews 33:548–557

Evans G, Axford D, Waterman D, Owen RL (2011) Macromolecular microcrystallography. Crystallogr Rev 17(2):105–142

J. R. Helliwell, *Certifying Central Facility Beamlines for Biological and Chemical Crystallography and Allied Methods*,
SpringerBriefs in Crystallography, https://doi.org/10.1007/978-3-031-80181-5_11

Hedman B, Hodgson KO, Helliwell JR, Liddington R, Papiz MZ (1985) Protein micro-crystal diffraction and the effects of radiation damage with ultra high flux synchrotron radiation. PNAS.USA 82:7604–7607

Flot D, Gordon EJ, Hall DR, Leonard GA, McCarthy A, McCarthy J, McSweeney S, Mitchell E, Nurizzo D, Ravelli RGB, Shepard W (2006) The care and nurture of undulator data sets. Acta Cryst D62:65–71

Hirata K, Foadi J, Evans G et al (2016) Structural biology with microfocus beamlines. In: Senda T, Maenaka K (eds) Advanced methods in structural biology. Springer Protocols Handbooks. Springer, Tokyo. https://doi.org/10.1007/978-4-431-56030-2_14

Chapter 12
Serial Crystallography Beamlines: Synchrotron

Abstract Stimulated by the XFELs introducing serial delivery of samples in liquid jets or extruded gels (see Chap. 13 below) synchrotron crystallography beamlines introduced similar approaches. [This section is placed here, ahead of the XFELS, to complete the synchrotron crystallography beamlines as a group of chapters.]

Stimulated by the XFELs introducing serial delivery of samples in liquid jets or extruded gels (see Chap. 13 below) synchrotron crystallography beamlines introduced similar approaches. [This section is placed here, ahead of the XFELS, to complete the synchrotron crystallography beamlines as a group of chapters.]

This type of beamline was introduced at EMBL Hamburg by Gati et al (2014) and elaborated on by Hakanpää et al (2018) and Stellato et al (2014), the former for microcrystals grown in vivo and the latter examples with a liquid jet of microcrystals.

The practical details for the fixed target method are excellently described in the article and accompanying six minutes video of Horrell et al (2021). This involves the use of micro-pipetted microcrystal slurries into a specially designed chip (i.e. samples holder). The video shows the preparation, loading and aligning the chip to the synchrotron X-ray beam as well as the use of the computer graphics user interface for the diffraction data indexing and integration. The authors provide specialised guidance for noticing cases of crystal polymorphs. They also emphasise dynamic crystallography studies of chemical reactions in the microcrystals initiated by light or rapid mixing methods. A comprehensive recent book including these methods is by Moffat and Lattman (2023).

References

Gati C, Bourenkov G, Klinge M, Rehders D, Stellato F, Oberthür D, Yefanov O, Sommer BP, Mogk S, Duszenko M, Betzel C, Schneider TR, Chapman HN, Redecke L (2014) Serial crystallography on in vivo grown microcrystals using synchrotron radiation. IUCrJ 1:87–94

Hakanpää J, Bourenkov G, Karpics I, Pompidor G, Bento I, Schneider T (2018) Serial synchrotron crystallography at EMBL PETRA III beamline P14. Acta Cryst. A 74:e172

Horrell S, Axford D, Devenish NE, Ebrahim A, Hough MA, Sherrell D A, Storm SLS, Tews I, Worrall JAR, Owen RL (2021) Fixed target serial data collection at diamond light source. J Vis Exp (168):e62200. https://doi.org/10.3791/62200

Moffat K, Lattman EF (2023) Dynamics and kinetics in structural biology: unravelling function through time-resolved structural analysis. Published by Wiley, New York, p 288. ISBN 978–1–119–69628–5

Stellato F, Oberthur D, Liang M, Bean R, Gati C, Yefanov O, Barty A, Burkhardt A, Fischer P, Galli L, Kirian RA, Meyer J, Panneerselvam S, Yoon CH, Chervinskii F, Speller E, White TA, Betzel C, Meents A, Chapman HN (2014) Room-temperature macromolecular serial crystallography using synchrotron radiation. IUCrJ 1:204–212

Chapter 13
Serial Crystallography Beamlines: XFELs

Abstract XFELs deliver a beam intensity in tens of femtoseconds that a typical synchrotron beamline delivers in a millisecond. The short intense pulse of the X-ray laser allows radiation damage mechanisms that occur on time scales longer than tens of femtoseconds to be avoided.

XFELs deliver a beam intensity in tens of femtoseconds that a typical synchrotron beamline delivers in a millisecond. The short intense pulse of the X-ray laser allows radiation damage mechanisms that occur on time scales longer than tens of femtoseconds to be avoided. This is known as the 'diffract before destroy' approach (Neutze et al. (2000)). Furthermore, the exceptional brightness of the XFEL reduces the required individual crystal sample volume by several orders of magnitude. [But n.b. there were huge increases in the total number of samples volume needed in an experiment though due to the uncertain chance of the XFEL beam to strike a crystal in the flow supply (see discussion below).] The X-ray flash destroys each tiny crystal that is hit due to the intense X-ray power. In addition, each diffraction pattern can be of limited statistical quality, that is, it has a low number of counts per diffraction spot. Therefore, many crystals are needed.

An early lucid description of serial crystallography with XFELs was provided by Chapman (2015). He described various methods used to deliver the sample into the X-ray beam. A liquid micro-jet as option was advocated as particularly flexible and that it gives a low X-ray background, suitable for sub-micron crystals. But as a sample delivery method it has a low sample usage efficiency, as mentioned above, because of its high speed, moving the liquid jet a relatively long distance between X-ray pulses so that many crystallites are simply missed by the XFEL pulses. Extrusion jets using such as a gel are slower, giving higher sample usage efficiency, but at the cost of higher X-ray background in the diffraction image.

The flowing systems operate in vacuum. Instead, a raster-scan approach of the tiny, focussed, X-ray pulse across a single crystal can move that crystal fast enough to hit every point giving optimum sample efficiency and data collection rate. This method could also be used on a random field of crystals. The single large crystal mounted on a goniometer can be exposed in the different places with specific angular increments between pulses. Sample delivery for XFEL experiments is an area in active development.

Schlichting (2015) provided a helpful overview of the achievements at XFELS in the first five years. A particularly challenging aspect has been the diffraction data processing software and procedures, and their optimisation for the new types of experiments. These experiments are intrinsically difficult, but best practices are emerging (Gorel et al (2021)).

An important demonstration described most recently by Moffat and Lattman (2024) (see their Chap.3 Box 3B discussion) was the comparison of synchrotron radiation time-resolved protein crystallography experiments with the ones done at the Stanford Linac Coherent Light Source (LCLS) whereby identical time points for the same biological system (photoactive yellow Protein, PYP) could be compared in a control experiment. Difference electron density signals were approximately doubled in the LCLS case. The advantage in the LCLS experiments over the synchrotron experiments was because of the much smaller crystals used and their better optical transparency to the laser pump thus establishing a higher occupancy of the structural intermediates. Moffat and Lattman (2024) also discuss anxieties about high peak powers of the laser pump initiating protein structure damaging multiphoton effects. They state: -

> An essential control experiment, a laser power titration should be carried and should reveal linear dependence of the X-ray signal on pump laser power from a typical sample….. Easy to state but in practice unrealistic to execute fully.

This systematic error in the method of photo crystallography is covered in detail by Barends et al (2024) including showing non-functional structural changes in CO bound to myoglobin. See also the associated News and Views article by Neutze and Miller (2024).

References

Barends TRM, Gorel A, Bhattacharyya S et al (2024) Influence of pump laser fluence on ultrafast myoglobin structural dynamics. Nature 626:905–911. https://doi.org/10.1038/s41586-024-070 32-9

Chapman HN (2015) Serial femtosecond crystallography. Synchrotron Radiation News 28(6):20–24. https://doi.org/10.1080/08940886.2015.1101323

Gorel A, Schlichting I, Barends TRM (2021) Discerning best practices in XFEL-based biological crystallography standards for nonstandard experiments. IUCrJ 8:532–543

Moffat K, Lattman EE (2024) Dynamics and kinetics in structural biology: unravelling function through time-resolved structural analysis. Published by Wiley, New York

Neutze R, Wouts R, van der Spoel D, Wecker E, Hajdu J (2000) Potential for biomolecular imaging with femtosecond X-ray pulses. Nature 406:752–757

Neutze R, Miller RJD (2024) Energetic laser pulses alter outcomes of X-ray studies of proteins. Nature 626:720–722

Schlichting I (2015) Serial femtosecond crystallography: the first five years. IUCrJ 2:246–255

Chapter 14
Time-Resolved-Crystallography and Uses Across Several Beamline Types

Abstract Crystallography has expanded into the time domain, which follows its success in yielding up very large numbers of 'static' crystal structures in both structural chemistry and then structural biology.

Crystallography has expanded into the time domain, which follows its success in yielding up very large numbers of 'static' crystal structures in both structural chemistry and then structural biology.

Some personal reminiscences are offered by the author of this Springer Brief:-

It is hard to pin down precisely when this started, but I think the invention of the protein crystallography flow cell at Yale University Molecular Biophysics seems to me to be a good starting point (Wycoff et al 1967). For chemical crystallography the pioneering work of Gerhard Schmidt seems to me a good starting point for that field (see e.g. Coppens and Schmidt 1965). In 1974 when I was interviewed for my DPhil position at the Laboratory of Molecular Biophysics in Oxford University by the Head, Professor David Phillips, he showed me a flow cell (very much like the Wyckoff et al (1967) one). These were conceived to initiate a reaction in an enzyme via diffusion of substrate into the crystal and using a lowering of the crystal sample temperature to slow down the enzyme reaction rate. During my DPhil I learnt that the X-ray protein crystallography data collection times of weeks or months for a data set on a rotating anode X-ray source with photographic film as detector had to be compared to an enzyme's functioning time in solution, typically in the milliseconds range but as short as microseconds or as long as seconds. Later as synchrotron beam line scientist at the UK's SRS 9.6 I could provide greatly reduced X-ray protein crystallography data collection times to support the pioneering studies of the glycogen phosphorylase enzyme "catalysis in the crystal" study (Hajdu et al (1987)).

A seminal contribution in the domain of physiologically relevant biological structure and function determination was by Keith Moffat, of Cornell and latterly of the University of Chicago, proposing that synchrotrons should offer the option of a Laue method data collection mode (Moffat, Szebenyi and Bilderback (1984)).

Today the field of time-resolved macromolecular crystallography has expanded enormously, reaching into the femtoseconds time domain using XFELs, matched to photosensitive proteins such as the oxygen evolving complex (for a recent overview see Bhowmick et al (2023)). The whole field spans time-resolved studies from that

J. R. Helliwell, *Certifying Central Facility Beamlines for Biological and Chemical Crystallography and Allied Methods*,
SpringerBriefs in Crystallography, https://doi.org/10.1007/978-3-031-80181-5_14

time range up to minutes (Caramello and Royant (2024) and, in the crystal, even hours and a year or two (Helliwell et al (1998), Brink and Helliwell (2019))).

Prior to the experiment at the beamline, it is important to characterise the dynamic system to be under study. At the ESRF the in crystallo optical spectroscopy (icOS) laboratory offers characterisation by Raman/UV fluorescence in situ and a "Cryo-bench "(Engilberge et al (2024)). These experiments provide complementary information to X-ray crystallography.

For the fast time-resolved structural crystallography any potential timing discrepancies between a light initiating pulse and the X-ray pulse for the crystallography will clearly be important. The LCLS at Stanford Linear Accelerator Center has a 'timing tool', described in detail here https://confluence.slac.stanford.edu/display/ PSDM/TimeTool. This is a good example of a critical certification of a beamline's technical diagnostic capabilities.

In chemical crystallography this field also includes "photo-crystallography", but which may be a study of metastable states i.e. not time resolved as such. Schmokel et al (2010) offer a way of calibrating the temperature rises in a crystal sample illuminated with multiple pulses of light. They do this from crystal sample unit-cell changes, analysis of what they call "photo-Wilson plots" and their comparison with "temperature-Wilson plots", and by refinement of a temperature scaling factor. All three of their methods agreed to within $\sim 10\%$. A very recent overview of photocrystallography is provided by Hatcher et al (2024).

References

Bhowmick A, Simon PS, Bogacz I, Hussein R, Zhang M, Makita H, Ibrahim M, Chatterjee R, Doyle MD, Cheah MH, Chernev P, Fuller FD, Fransson T, Alonso-Mori R, Brewster AS, Sauter NK, Bergmann U, Dobbek H, Zouni A, Messinger J, Kern J, Yachandra VK, Yano J (2023) Going around the Kok cycle of the water oxidation reaction with femtosecond X-ray crystallography. IUCrJ 10:642–655

Brink A, Helliwell JR (2019) Formation of a highly dense tetra-rhenium cluster in a protein crystal and its implications in medical imaging. IUCrJ 6:695–702

Caramello N, Royant A (2024) From femtoseconds to minutes: time-resolved macromolecular crystallography at XFELs and synchrotrons. Acta Cryst D80:60–79

Coppens P, Schmidt GMJ (1965) The crystal structure of the metastable (β) modification of p-nitrophenol. Acta Cryst. 18:654–663

Engilberge S, Caramello N, Bukhdruker S, Byrdin M, Giraud T, Jacquet P, Scortani D, Biv R, Gonzalez H, Broquet A, van der Linden P, Rose SL, Flot D, Balandin T, Gordeliy V, Lahey-Rudolph JM, Roessle M, de Sanctis D, Leonard GA, Mueller-Dieckmann C, Royant A (2024) The TR-icOS setup at the ESRF: time-resolved microsecond UV/Vis absorption spectroscopy on protein crystals. Acta Cryst D80:16–25

Hajdu J, Acharya KR, Stuart DI, McLaughlin PJ, Barford D, Oikonomakos NG, Klein HW, Johnson., L.N. (1987) Catalysis in the Crystal: Synchrotron Radiation Studies with Glycogen Phosphorylase b. EMBO J 6:539–546

Hatcher LE, Warren MR, Raithby PR (2024) Methods in molecular photocrystallography. Acta Cryst C80:585–600

Helliwell JR, Nieh YP, Raftery J, Cassetta A, Habash J, Carr PD, Ursby T, Wulff M, Thompson AW, Niemann AC, Hädener A (1998) Time–resolved structures of hydroxymethylbilane synthase (Lys59Gln mutant) as it is loaded with substrate in the crystal determined by Laue diffraction. Faraday Trans 94(17):2615–2622

Moffat K, Szebenyi D, Bilderback D (1984) D. X-ray Laue Diffraction from Protein Crystals. Science 223(4643):1423–5

Schmokel MS, Kaminski R, Benedict JB, Coppens P (2010) Data scaling and temperature calibration in time-resolved photo-crystallographic experiments. Acta Cryst A66:632–636

Wyckoff HW, Doscher M, Tsernoglou D, Inagami T, Johnson LN, Hardman KD, Allewell NM, Kelly DM, Richards FM (1967) Design of a diffractometer and flow cell system for X-ray analysis of crystalline proteins with applications to the crystal chemistry of ribonuclease-S. J Mol Biol 27:563–578

Chapter 15
Electron Crystallography Facilities

Abstract The stronger scattering strength of electrons than X-rays by approximately a million times has led to a revolution in single particle cryoEM (see Chap. 20). But also, electron nanocrystallography has seen a steady evolution for some 50 years.

The stronger scattering strength of electrons than X-rays by approximately a million times has led to a revolution in single particle cryoEM (see Chap. 20). But also, electron nanocrystallography has seen a steady evolution for some 50 years (Vainshstein (1964) and Zou et al (2011)) but its expansion into macromolecular crystallography these past 10 years or more has been impressive. Hence, difficult to grow membrane protein crystals, when in the micron size range, have been successfully tackled (Clabbers et al (2022)).

An article and accompanying video by Martynowycz and Gonen (2021) describe the practical details of the method for nano and micro sized crystals. They give a case study of a dataset for carbamazepine.

References

Clabbers MTB, Shiriaeva A, Gonen T (2022) MicroED: conception, practice, and future opportunities. IUCrJ 9:169–179

Martynowycz MW, Gonen T (2021) Microcrystal electron diffraction of small molecules. J vis Exp 169:e62313. https://doi.org/10.3791/62313

Vainshtein B (1964) Structure analysis by electron diffraction. Pergamon Press, Oxford

Zou X, Hovmoller S, Oleynikov P (2011) Electron crystallography: electron microscopy an electron diffraction IUCr texts on crystallography. Oxford University Press, Oxford

Chapter 16
Neutron Crystallography Instruments

Neutrons have a unique role to play in determining the structure and dynamics of biological macromolecules and their complexes. The similar neutron scattering magnitude from deuterium (^{2}H), carbon (C), nitrogen (N), and oxygen (O) nuclei, means that the effect of atomic vibration in lowering the visibility of these atoms in Fourier maps is no worse for the deuterium atoms, unlike the X-ray case. Moreover, the negative scattering length of the common protium isotope (^{1}H) and the positive scattering length of deuterium allows the well-known neutron ^{1}H/^{2}H contrast variation method to be applied. Furthermore, as there is no problem with radiation damage using neutrons as the diffraction probe, unlike X-rays, room temperature neutron data collection studies are completely viable. Unfortunately, even with these advantages, the low flux of existing neutron facilities means that neutron biological crystallography is not going to be a high throughput technique due to the long measuring runs at present e.g. typically between one to two weeks or more.

The importance of this method rests on needing to know the details of the hydrogen atoms and bound water structure, which are involved in virtually all the molecular processes of life. This experimental structural information is incomplete when studied by X-rays alone. Also, many enzyme reactions involve hydrogens. Consequently, approximately 70% of all neutron structures deposited in the PDB are enzymes. Thus, there is great potential for even wider application, should the technical capability be found. There are two major hurdles for wide application; firstly, the size of crystals routinely available versus the sizes required (typically ~ 1 mm^3), and secondly a molecular weight ceiling limit of typically 40 kDa.

Current and planned instruments are given in Table 16.1. Also, routine use of full deuteration of the protein through microbiological expression of proteins for bacteria

---

The original version of the chapter has been revised. A correction to this chapter can be found at https://doi.org/10.1007/978-3-031-80181-5_31

J. R. Helliwell, *Certifying Central Facility Beamlines for Biological and Chemical Crystallography and Allied Methods*, SpringerBriefs in Crystallography, https://doi.org/10.1007/978-3-031-80181-5_16

Table 16.1 Neutron sources with facilities for neutron biological crystallography around the world

Country	Facility	Stations	General webpage
Australia	ANSTO, Sidney	KOALA[a]	https://www.ansto.gov.au/our-facilities/australian-centre-for-neutron-scattering/neutron-scattering-instruments/koala-laue
France	ILL, Grenoble	LADI[a] DALI[ca]	http://www.ill.fr/YellowBook/LADI/
Germany	FRM2, Munich	BioDiff	https://mlz-garching.de/biodiff
Japan	J-PARC, Ibaraki	BL-03[a]	https://j-parc.jp/researcher/MatLife/en/applying_list/list2009A.html
Sweden	ESS, Lund	NMX[c]	https://europeanspallationsource.se/instruments/nmx
United Kingdom	ISIS TS2, Oxford	LMX[b]	https://www.isis.stfc.ac.uk/Pages/Endeavour-LMX.aspx
USA	HIFER, Tennessee	IMAGINE[a]	https://neutrons.ornl.gov/imagine
USA	SNS, Tennessee	MANDI[a];	https://neutrons.ornl.gov/mandi

[a] Currently operating
[b] Funding obtained in 2023
[C] Under construction

grown on deuterated media is usually done at all these research facility centres. This increases the crystal sample scattering strength for neutrons and allows a smaller crystal volume to be used. This advantage is in addition to avoiding the nuclear density cancellation effect of a hydrogen bound to a nitrogen or carbon atom.

A review of the method is provided by Blakeley (2009). See also the textbook by Niimura and Podjarny (2011) and the edited compilation led by Moody (2020). The practical details are excellently described in the article and accompanying ten minutes video of Schröder and Meilleur (2020).

References

Blakeley MP (2009) Neutron macromolecular crystallography. Crystallogr Rev 15(3):157–218
Moody, P.C.E. (2020) Neutron crystallography in structural biology methods in enzymology, vol 634, 2-389
Niimura N, Podjarny A (2011) Neutron protein crystallography: hydrogen, protons, and hydration in bio-macromolecules. OUP, Oxford
Schröder GC, Meilleur F (2020) Neutron crystallography data collection and processing for modelling hydrogen atoms in protein structures. J Vis Exp 166:e61903. https://doi.org/10.3791/61903

Chapter 17
Facility Data Archiving Policy Considerations

The evolution of photon and neutron central facilities' data policies in the past two decades or so has seen substantial changes. These changes have reflected the practicality of being able to archive large i.e. '*Big Data*' level of quantities. However, there has also, first, been an evolution of policy thinking especially by the funding agencies as they increasingly realised that commercial publishers were making large profits (even around 50%) out of taxpayers' funded research. Clearly this was a violation of principle, not least that a member of that tax paying public could not access the research results in scientific journals that the funding agencies as their proxies had funded. An awkward point in this simple and obviously compelling argument was that the funding agencies typically funded 'only' about 20% or so of the proposals put to them. So, what about the results from unfunded research? A second aspect was that many of the learned societies had their own journals which made only small surpluses which in any case was invested in schemes like student bursaries for their training. Nevertheless, an unstoppable momentum has built up in ensuring open access to all research results.

That research results should be presented along with their underpinning data has been a tradition of crystallographers, introduced by Bragg (1913) and formalised in the crystallographic databases firstly with the Cambridge Structural Database launched in 1965 and the PDB launched in 1971. A wide spectrum of databases is available today as summarised in Bruno et al. (2017).

A landmark in policy development was the report of the Organisation for Economic Co-operation and Development (OECD) (2007) within which it states they wished to: -

---

The original version of the chapter has been revised. A correction to this chapter can be found at https://doi.org/10.1007/978-3-031-80181-5_31

J. R. Helliwell, *Certifying Central Facility Beamlines for Biological and Chemical Crystallography and Allied Methods*, SpringerBriefs in Crystallography, https://doi.org/10.1007/978-3-031-80181-5_17

provide broad policy recommendations to the governmental science policy and funding bodies of member countries on access to research data from public funding. They are intended to promote data access and sharing among researchers, research institutions, and national research agencies, while at the same time, recognising and taking into account, the various national laws, research policies and organisational structures of member countries. The ultimate goal of these Principles and Guidelines is to improve the efficiency and effectiveness of the global science system.

The OECD (2007) also defined "research data":

as factual records (numerical scores, textual records, images and sounds) used as primary sources for scientific research, and that are commonly accepted in the scientific community as necessary to validate research findings....(and) are principally aimed at research data in digital, computer-readable format.

In the section on Professionalism, it states that: -

In current research practice, the initial data-producing researcher or institution is sometimes rewarded with temporary exclusive use of the data. The rules for such incentive arrangements should be developed and explicitly stated by the funding sources in co-operation with the affected research communities.

In the final section on Sustainability, it states that:

(sustainability) can be a difficult task, given that most research projects, and the public funding provided, have a limited duration, whereas ensuring access to the data produced is a long-term undertaking.

These guidelines from the OECD presented a challenge, as well as opportunities, to all universities, principal investigators, and the central facilities. It also challenged the funding agencies who had to give a clear (or clearer) budget line to data management, storage, and access costs. Prior to 2000, approximately, the UK's Daresbury SRS, where JRH had worked since its inception in 1981 including as Director, had a data policy that was basically "*measured data will be retained by the facility only for 28 days after beamtime*". Thereafter it was the responsibility of the user to transport their raw data home, take their analyses forward and communicate the results. The SRS would then assemble an annual report where publications from the use of the SRS would be listed as an Appendix to the thematic summaries written by in house staff in consultation with some selected users. The impact of the investment in the SRS and its experimental programme would then be the yearly number of those publications and the highlighted studies such as in high impact journals and securing front covers of journals. Highlights in the media such as via newspapers and television could also be noted in these SRS facility annual reports.

Beamtime proposals at the outset of a study would have had to state that the work was not being undertaken elsewhere such as at another synchrotron. It was obvious to the in-house staff that this was not adhered to as beamtime was scarce, and there were regular chances for in house staff to compare notes with colleagues at other facilities. It was also suspected that not all the data measured from beamtime runs led to a publication. There was no means to investigate these concerns. However, there was steadily an increase in digital data archive capacities, and which brought

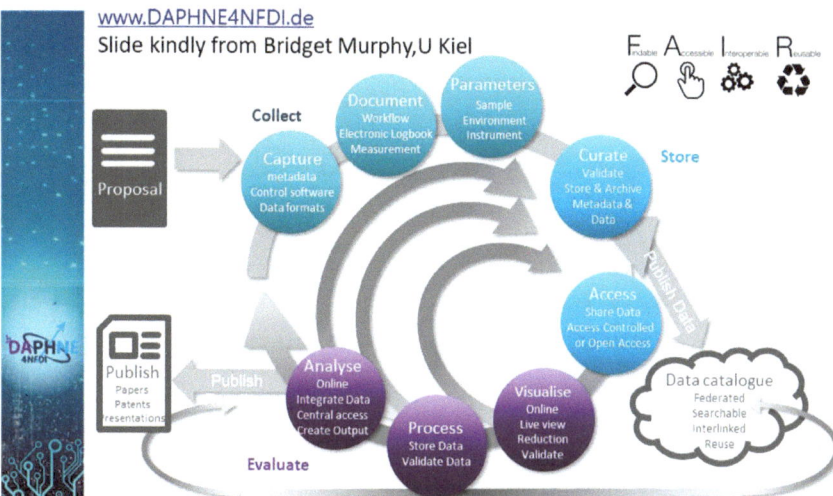

Fig. 17.1 DAPHNE4NFDI Research data management strategy envisaged for the central facilities in Germany. https://www.daphne4nfdi.de/english/index.php Figure kindly provided by and reproduced here with the permission of Professor Dr Bridget Murphy, University of Kiel, Germany and Co-Director of the DAPHNE4NFDI Consortium; the diagram's source is the European Union's "ExPaNDS/PaNOSC H2020 Projects"

with them the chance to shift the responsibilities for preserving the measurements from the user to professional data managers and IT specialists at the central facility. Users by and large welcomed this change. Data policies at the facilities could then evolve to reflect these changes. A common theme emerged, in mainland Europe at least, consistent with the OECD (2007) guiding principles, that: -

measured data will be retained by a facility for at least ten years

and

measured data will be made public after three years.

Furthermore, there is a concept that has developed that there should be an integrated life cycle of facility proposal, then data measurement including electronic laboratory notebook metadata recording, data archiving at the facility including made open after 3 years, then linking to publication; see e.g. Fig. 17.1.

Overall, the crystallography community has developed over many years a coherent framework for its information and metadata/data (Fig. 17.2).

Examples of sample and experimental metadata sought at a home laboratory for both chemical X-ray crystallography and powder X-ray diffraction are shown in Fig. 17.3. The metadata about the beamline is carried in the raw diffraction images header to the file.

There are caveats to emphasise. Firstly, globally, facility data archiving policies vary by region. In the USA, for example, the facilities still principally deem the

Overall, there is a coherent approach of crystallography:
Crystallographic Information Framework (CIF) ontologies at each stage

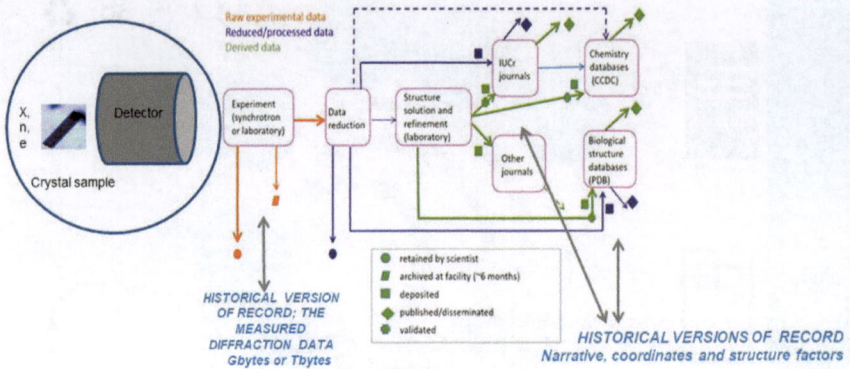

Fig. 17.2   A coherent information flow in crystallography. CIF ontologies characterize data at every stage of the information processing life cycle, from experimental apparatus to published paper and curated database deposit. Figure based on Kroon-Batenburg et al (2017)

data the users' property and responsibility. Secondly, there are variations between types of facility i.e. for the XFEL and EBS facilities the data loads, both rates and volumes, is very challenging to manage. The EuXFEL has recently amended its policy accordingly. The relevant EuXFEL documents can be found at:

https://www.xfel.eu/sites/sites_custom/...23_eng.pdf
https://www.xfel.eu/sites/sites_custom/...ft_eng.pdf

The EuXFEL is balancing opportunities and cost of raw data archiving, as well as ensuring reproducibility and giving access to historic measurements for new software to the best extent possible. Overall, global crystallography community policies are described in the following recent publications: Brink et al. (2024), Hackert et al. (2016), Helliwell et al. (2019), International Science Council (2015), Kroon-Batenburg et al. (2024).

UNIVERSITY OF THE FREE STATE
UNIVERSITEIT VAN DIE VRYSTAAT
YUNIVESITHI YA FREISTATA

UFS
NATURAL AND
AGRICULTURAL SCIENCES

Application for single crystal data collection

Applicant details

Name							
Crystal code		Sample queue position		Data backup requirements		CD / DVD	
						Network	
						e-mail	
Proposed structure							
Empirical formula							

Crystal data of CCDC hits (starting materials, by-products, etc)	Compound	Compound	Compound
	a	a	a
	b	b	b
	c	c	c
	α	α	α
	β	β	β
	γ	γ	γ
	Bravais	Bravais	Bravais

Solvents	Act	DCM	CHCl₃	MeOH	Tol	CH₃CN	H₂O	Eth	Other	
Additional Characterisation		NMR	IR	Element analysis		Mass spec	Other			
Comments										

Data collection details

Code				Collection date		/	/ 20
Scans	φ		ω	Nr. of frames			
Frame width			°	Detector Distance			cm
Total time			h	Exposure time			s
Temperature			K	Radiation used		Mo	Cu
Crystal size	X	X	mm³	Color			
Absorption correction	none	multi-scan	face-index	Structure solved		Yes	No
Comments				Completeness			
				Archive			

Fig. 17.3 Sample and experimental metadata logging is vital for any attempt to redo or reanalyse any measured data. Top: this example is for the chemical crystallography facility. Bottom: this example is for powder X-ray diffraction. Both these forms are used at the University of the Free State, South Africa, Chemistry Department kindly provided by Dr. Alice Brink

Department of Chemistry

UNIVERSITY OF THE FREE STATE
UNIVERSITEIT VAN DIE VRYSTAAT
YUNIVESITHI YA FREISTATA
UFS·UV
NATUURAL AND AGRICULTURAL SCIENCES
NATUUR- EN LANDBOUWETENSKAPPE

Application for PXRD 2024 data collection (Client)

Surname		University/ Company name		Mentor/ Supervisor	
Student/Staff No.					
				Email	
Email		Department			
Sample Description (eg. Black powder, yellow crystals, etc.)		Chemical composition (eg. Si, Zn, etc)		Analysis Theta range (eg.5°-90°)	
Number of samples					
Analysis (tick correct box)					
	PXRD Data		Data Processing		Phase Identification Percentage Crystallinity Crystal size
Data Collection details (Analyst)					
Scan type		Theta range (degree theta)		Scan mode	
Time (s) /Steps		Radiation source		Detector	VÅNTEC-1
Current (mA)		Voltage (KV)			

205 Nelson Mandela Drive/Rylaan, Park West/Parkwes, Bloemfontein 9301, South Africa/Suid-Afrika
P.O. Box/Posbus 339, Bloemfontein 9300, South Africa/Suid-Afrika, T: +27(0)51 401 2439, www.ufs.ac.za

Fig. 17.3 (continued)

References

Bragg WL (1913) Proc R Soc Lond A 89:248–277

Brink A, Bruno I, Helliwell JR, McMahon B (2024) The interoperability of crystallographic data and databases. IUCrJ 11:9–15

Bruno I, Gražulis S, Helliwell JR, Kabekkodu SN, McMahon B, Westbrook J (2017) Crystallography and databases. Data Sci J 16:38–38

Hackert ML, van Meervelt L, Helliwell JR, McMahon B (2016) Open data in a big data world: a position paper for crystallography. https://www.iucr.org/iucr/open-data

Helliwell JR, Minor W, Weiss MS, Garman EF, Read R J, Newman J, van Raaij MJ, Hajdu J, Baker EN (2019) Findable accessible interoperable re-usable (FAIR) diffraction data are coming to protein crystallography. Acta Cryst D75:455–457

International Science Council (2015). https://council.science/publications/open-data-in-a-big-data-world

Kroon-Batenburg LMJ, Helliwell JR, McMahon B, Terwilliger TC (2017) Raw diffraction data preservation and reuse: overview, update on practicalities and metadata requirements. IUCrJ 4:87–99

Kroon-Batenburg LMJ, Lightfoot MP, Johnson NT, Helliwell JR (2024) Raw diffraction data and reproducibility. Struct Dyn 11(1):011301

OECD (2007) OECD principles and guidelines for access to research data from public funding. https://www.oecd.org/science/inno/38500813.pdf. Organisation for Economic Co-operation and Development

Chapter 18
Facility Publication Authorship Policies

Starting with a reminiscence:

> When I was running the protein crystallography beamlines at the UK's SRS in the 1980s, I was approached for a meeting on this topic of facility publication with my equivalents at two other European SR facilities, EMBL Hamburg and LURE DCI in Paris. The point at issue was whether every user beamtime run that led to publication should have beamline staff as coauthors and us in particular. I resisted this proposal for several reasons but top of my list was that there were competing groups in the 'big topics' and if they got into a dog fight with their articles having different conclusions it would not sit well on me as a coauthor of both, or more, such papers. I offered the alternative view that we each ensure our beamlines were described in a publication, and that would be required as a citation in any results paper from our respective beamlines. This was agreed with the exception where the work was obviously a collaboration, and our role was substantive. This last word 'substantive' is rather difficult to define well. A senior colleague in the field independently remarked to me that "some user runs involve measurements and other runs involve experiments". I took the point to mean that 'measurements' involved a routine use of the beamline whereas 'experiments' were not routine which suggests that coauthorship of beamline staff would then likely be appropriate.

> These points served me well I felt apart from two situations: firstly that many users were not necessarily scrupulous in citing one's beamline and secondly that on some occasions I did feel my role was substantive, warranting co-authorship, but this was not given to me. Today, like data management (and data sharing), there should be a publication policy before any experiment is undertaken. These policies are now, to my knowledge, available at all central facilities and thereby clear to users and in house staff alike.

As a modern example, let's refer to the ESRF, which as a pan European project involves at least 10 participating countries with national and EU laws to navigate and whose policy will be an amalgamation of those countries' national laws. The ESRF publications and data policies can be found here https://www.esrf.fr/home/UsersAndScience/esrf-user-policies-and-rules.html. This offers very clear statements, and one sentence has underlined portions:

> Before using beamtime, all users electronically validate and sign a statement (in the User & Safety Declaration form) that they will give proper credit to ESRF staff members and experimental facilities, the beamline in particular.

© The Author(s), under exclusive license to Springer Nature Switzerland AG 2025
J. R. Helliwell, *Certifying Central Facility Beamlines for Biological and Chemical Crystallography and Allied Methods*,
SpringerBriefs in Crystallography, https://doi.org/10.1007/978-3-031-80181-5_18

55

These ESRF guidelines go on to describe what precisely is 'proper credit'. Several circumstances are described where each one of these could warrant co-authorship. The Guidelines also state that *"Scientists purchasing proprietary research beamtime are not obliged to publish results, but if they do so they must respect the obligations (described for non-proprietary research)."*

Not all measured data lead to a publication. This can be for several reasons such as:

- Analysis is found to be beyond the scope of current knowledge or software available at the time.
- The Principal Investigator (PI) for the research is unable to give sufficient priority to writing a publication or is suffering a period of ill health.

What is the best option in these non-publication eventualities? Overall, a lack of publication in a timely manner for each user run, or collection of themed runs, is to the detriment of the taxpayers of the participating countries who have funded the facility and quite probably the PI's team. The *respect for the taxpayer* is the principle that drives the policy-engine that is 'open science'. So, in the absence of a publication after three years, the European view is that the default policy should be release to the public i.e. 'open data' after three years, known as the embargo period. An appeals' procedure exists for the PI to extend that embargo period. Precisely what constitutes 'good reasons' for the Facility Directors to grant the Appeal from a PI are not yet clear. To allow for the possibility that a PI may become ill a prior nomination of a Deputy is needed.

An insufficient weight is given on the release of unpublished measurements to the viewpoint of the PI who I think is usually the best judge of when it is timely to share results and data via a publication. IUCr Journals have launched Raw Data Letters (https://www.iucr.org/news/newsletter/volume-30/number-3/iucrdata-launches-raw-data-letters) as a new article category for PIs to share raw data that e.g. present analysis challenges that might be solved by open sharing of raw data. Poor quality data however may simply be better to be remeasured e.g. under more stringent experimental conditions in a future user run.

It is not only at the central facilities where these issues must be addressed. Small molecule crystallographers have also felt that credit was not necessarily given to them especially if they were employed as technical officers at a university as "service crystallographers", i.e. the problem word in that work descriptor is "service". The Cambridge Crystallographic Data Centre (CCDC) has helped mitigate this by requiring that the crystallographer responsible for data collection and refinement must be identified during data deposition into their Crystal Structure Database (the CSD).

Chapter 19
Sample Environment, Cryostats and or Containment, Light Sources for Photo-Initiation or Mixing Devices for Reactions in a Crystal

The most commonly available sample condition used is that of a cryo-temperature. Cooling a crystal in smaller molecule crystallography for diffraction data collection at around 100 K is the norm. In biological crystallography where X-ray radiation damage at ambient temperature is a problem this challenge is routinely 'solved' by cooling in an acceptable manner the crystal using a cryogen such as liquid nitrogen. Best practice for preventing a too big an increase in a crystal's mosaicity on cooling is described by Mitchell and Garman (1994); see especially their Fig. 2 re the selection of a cryoprotectant percentage. Warkentin and Thorne (2009) present results that demonstrate that high-quality diffraction data can be continuously collected from an individual crystal, over a temperature range between 300 and 100 K, by slow cooling at 0.1 K s^{-1}.

A very common type of experiment using cryostats is magnetic crystallography or superconductivity studies using neutrons. These are described here, as an example, of what is available at a leading neutron research facility: https://www.ill.eu/users/support-labs-infrastructure/sample-environment/services-for-advanced-neutron-environments/history/cryogenics/orange-cryostats/.

Light sources for photo crystallography are diverse. Brayshaw et al. (2010) describe use of a light emitting diode and a very recent overview is by Hatcher et al (2024). Coppens et al (2017) recommend a priori checking of the light-response and data quality before extended data collection in pump-probe photo crystallography experiments. The XFELs undertake time-resolved structural studies into the femtosecond range and will thereby include the need for the pumping laser to also have a pulse capability into the same time-resolution. As an example, the EuXFEL laser pump-probe facilities are described by Palmer et al (2019).

Sometimes crystals or fibres are held at a particular humidity. Most famously probably is the hydration of DNA fibres being important to select particular structural states. Rosalind Franklin did this (Franklin and Gosling (1953)), and which led to her "DNA X-ray photograph 57". Kiefersauer et al (2000) described a novel device for capillary-free mounting of protein crystals, based on surrounding the crystal with an

J. R. Helliwell, *Certifying Central Facility Beamlines for Biological and Chemical Crystallography and Allied Methods*, SpringerBriefs in Crystallography, https://doi.org/10.1007/978-3-031-80181-5_19

air stream of controlled humidity. Sanchez-Weatherby et al (2009) developed a user-friendly apparatus for dehydration experiments capable of generating a controlled humid airstream whilst allowing full functionality on most X-ray crystallography beamlines for it to be widely and routinely used by the MX community. Bowler et al. (2017) provide a theoretical description of such humidity control sufficient to allow the accurate prediction of humid atmospheres for specific sample requirements.

References

Bowler MG, Bowler DR, Bowler MW (2017) Raoult's law revisited: accurately predicting equilibrium relative humidity points for humidity control experiments. J Appl Cryst 50:631–638

Brayshaw SK, Knight JW, Raithby PR, Savarese TL, Schiffers S, Teat SJ, Warren JE, Warren MR (2010) Photo crystallography—design and methodology for the use of a light-emitting diode device. J Appl Cryst 43:337–340

Coppens P, Makal A, Fournier B, Jarzembska KN, Kaminski R, Basuroy K, Trzop E (2017) A priori checking of the light-response and data quality before extended data collection in pump-probe photocrystallography experiments. Acta Cryst B73:23–26

Franklin RE, Gosling RG (1953) The structure of sodium thymonucleate fibres. I. The influence of water content. Acta Cryst 6:673–677

Hatcher LE, Warren MR, Raithby PR (2024) Methods in molecular photocrystallography. Acta Cryst C80:585–600

Kiefersauer R, Than ME, Dobbek H, Gremer L, Melero M, Strobl S, Dias JM, Soulimane T, Huber R (2000) A novel free-mounting system for protein crystals: transformation and improvement of diffraction power by accurately controlled humidity changes. J Appl Cryst 33:1223–1230

Mitchell EP, Garman EF (1994) Flash freezing of protein crystals: investigation of mosaic spread and diffraction limit with variation of cryoprotectant concentration. J Appl Cryst 27:1070–1074

Palmer G, Kellert M, Wang J, Emons M, Wegner U, Kane D, Pallas F, Jezynski T, Venkatesan S, Rompotis D, Brambrink E, Monoszlai B, Jiang M, Meier J, Kruse K, Pergament M, Lederer MJ (2019) Pump-probe laser system at the FXE and SPB/SFX instruments of the European X-ray free-electron laser facility. J Synchrotron Rad 26:328–332

Sanchez-Weatherby J, Bowler MW, Huet J, Gobbo A, Felisaz F, Lavault B, Moya R, Kadlec J, Ravelli RBG, Cipriani F (2009) Improving diffraction by humidity control: a novel device compatible with X-ray beamlines. Acta Cryst D65:1237–1246

Warkentin M, Thorne RE (2009) Slow cooling of protein crystals. J Appl Cryst 42:944–952

Chapter 20
Electron Biological Imaging Centres at the SR Facilities: Diamond, ESRF, Soleil and SSRL/SLAC as Examples

Let me start with a reminiscence: -

> I find it a wonderful instrument, an electron microscope. As an undergraduate physics student in the early 1970s at York University, UK I undertook a final year project on the 'Determination of the Burgers vectors of dislocations in thin metal films'. The quality of the thin metal film as an ordered array of metal atoms was of course immediately visible via the diffraction mode. One could then switch to imaging mode if the metal film was of adequate quality to see the dislocations clearly. The imaging mode is not available of course using X-rays or neutrons as probes because suitable lenses are not available to focus them after transmission through a sample, at least at the atomic resolution usually required for understanding structure, dynamics and mechanism. In any case with electron biological imaging single particles of a biological macromolecule or complex are imaged.

As particle preservation techniques at cryo-temperatures improved, as well as the sensitivity of detectors, and their time-resolution, to the electrons transmitted through the single particles improved, the spatial resolution achievable with electron microscopy took a revolutionary leap to the atomic level. This was about ten years ago. There is an interesting restriction that the macromolecules or their complexes need to be a certain size such as above 100 kDa, depending on the spatial resolution achieved. Thus, a complementarity with macromolecular crystallography exists. Secondly though macromolecular complexes often defy crystallisation, with major exceptions like viruses or the ribosome. Even with these last two categories today it is generally true that cryoEM yields a structure determination more rapidly and with less material than crystallography does. Also, a cryoEM single particle analysis allows the pooling into subsets of the measured images so that different conformations and/or arrangements can be seen giving insight into the flexibilities of the complex to be seen.

Synchrotron facilities with their extensive experience with serving user communities launched electron bioimaging centres adjacent to their synchrotron. A pioneer in this was the Diamond Light Source. This was followed by others such as ESRF, Soleil and SSRL/SLAC, as examples. Table 20.1 provides weblinks and a short description

© The Author(s), under exclusive license to Springer Nature Switzerland AG 2025
J. R. Helliwell, *Certifying Central Facility Beamlines for Biological and Chemical Crystallography and Allied Methods*,
SpringerBriefs in Crystallography, https://doi.org/10.1007/978-3-031-80181-5_20

Table 20.1 Electron bioimaging centres around the world at the synchrotron facilities

Country	Synchrotron facility	Microscopes by theme	General webpage
France	ESRF EBS with EMBL Grenoble	CryoEM	https://www.esrf.fr/CM01
France	Soleil	CryoEM	Website under construction
Japan	SPring-8	CryoEM	https://user.spring8.or.jp/?p=37403&lang=en
Spain	ALBA	CryoEM	https://www.ibmb.csic.es/en/platforms/cryo-electron-microscope/
UK	Diamond Light Source	Single particle cryoEM; Cryo-electron tomography; MicroED; CryoFIB/SEM Lamellae; Cryo-PFIB-SEM volume imaging; Correlative Light and Electron Tomography	https://www.diamond.ac.uk/Instruments/Biological-Cryo-Imaging/eBIC/Capabilities.html
USA	SSRL	CryoEM	https://cryoem.slac.stanford.edu/
USA	NSLS II	CryoEM	https://www.bnl.gov/cryo-em/

Nb there are other cryoEM service centres, or which have offshoots of research centres who do some service work; these are not documented here

of what is available at each centre. Note that some of these facilities also include such as microED (see Chap. 15) and cryoelectron tomography for in cell imaging. Details of these are also given in Table 20.1. There may well be prior experiments to be done to qualify for the use of the electron microscope services; an example of this is shown in Fig. 20.1 from the ESRF service (Kandiah et al 2019).

The validation of cryoEM studies is maturing. The most recent update is from Kleywegt et al. (2024). Also, the capabilities of electron microscopes, and their detectors, are still improving. So, such as the lower molecular weight restriction, which I quote above as around 100 kDa, is lowering. There are examples that single particle cryo-EM is good to about 70 kDa and with a tag on the molecule of interest, e.g. an antibody, can go much smaller.

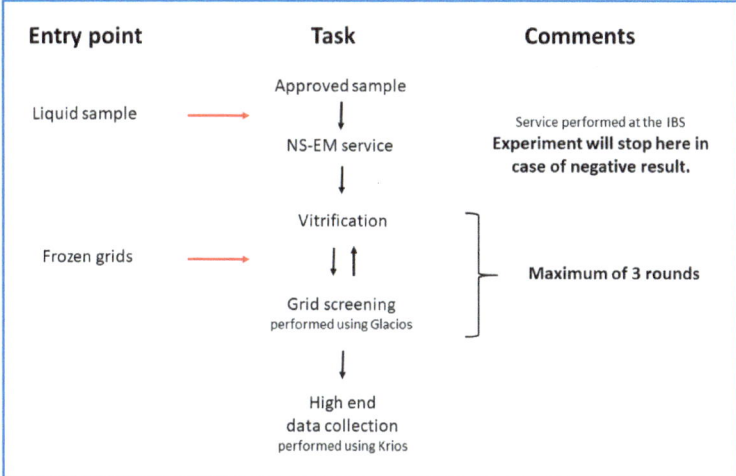

Fig. 20.1 The Solution to Structure (SOS) Service provided at the ESRF. An "approved liquid sample of a macromolecular assembly" is described as having been characterised by "gel filtration profile, multi-angle light scattering, dynamic light scattering, small angle X-ray scattering etc." or from frozen cryoEM grids that are not yet screened. NS-EM is negative stain electron microscopy. Reproduced with the permission of the ESRF

References

Kandiah E, Giraud T, de Maria Antolinos A, Dobias F, Effantin G, Flot D, Hons M, Schoehn G, Susini J, Svensson O, Leonard GA, Mueller-Dieckmann C (2019) CM01: a facility for cryo-electron microscopy at the European Synchrotron. Acta Cryst D75:528–535

Kleywegt GJ, Adams PD, Butcher SJ, Lawson CL, Rohou A, Rosenthal PB, Subramaniam S, Topf M, Abbott S, Baldwin PR, Berrisford JM, Bricogne G, Choudhary P, Croll TI, Danev R, Ganesan SJ, Grant T, Gutmanas A, Henderson R, Heymann JB, Huiskonen JT, Istrate A, Kato T, Lander GC, Lok S.-M, Ludtke SJ, Murshudov GN, Pye R, Pintilie GD, Richardson JS, Sachse C, Salih O, Scheres SHW, Schroeder GF, Sorzano COS, Stagg SM, Wang Z, Warshamanage R, Westbrook JD, Winn MD, Young JY, Burley SK, Hoch JC, Kurisu G, Morris K, Patwardhan A, Velankar S (2024) Community recommendations on cryoEM data archiving and validation. IUCrJ 11 (In press)

Chapter 21
NMR Crystallography

A major treatise on this topic is by Harris et al. (2009). This book describes the whole field, which is very broad. NMR responds to the short-range environment of relevant atoms and is not directly influenced by long-range order. It can therefore be applied to amorphous materials as well as crystalline ones and the solution state.

NMR can determine the chemical nature of a solid compound, including crystallographically important information such as conformation and tautomeric form. NMR chemical shifts give information about inter- and intra-molecular hydrogen-bond linkages. Polymorphs are usually easily distinguished. Phase transitions can be monitored. Crystallographic disorder is detectable, and distinctions between spatial and temporal disorder can be made. Motions such as internal rotation and ring inversion can be detected and their rates obtained, even in cases of mutual exchange (e.g. 180° ring flips of phenyl groups). NMR data can be used as restraints in carrying out full structure determination from powder diffraction data. In the case of host–guest complexes the guest can be mobile and obviously thereby disordered and its visibility absent or limited in the diffraction-derived structural analyses. These can be understandable from the NMR measurements.

The combined use of solution NMR and X-ray crystallography in structural biology is described by Rinaldelli et al (2014).

NMR spectrometers are usually available at individual university departments and are too numerous to summarise here. A Commission on NMR crystallography and Related Methods has been established by the International Union of Crystallography and is described here (https://www.iucr.org/iucr/commissions/nmr-crystallography).

References

Harris RK, Wasylishen RE, Duer MJ (eds) (2009) NMR crystallography. Wiley, Chichester, 504pp. ISBN 978-0-470-69961-4

© The Author(s), under exclusive license to Springer Nature Switzerland AG 2025
J. R. Helliwell, *Certifying Central Facility Beamlines for Biological and Chemical Crystallography and Allied Methods*,
SpringerBriefs in Crystallography, https://doi.org/10.1007/978-3-031-80181-5_21

Rinaldelli M, Ravera E, Calderone V, Parigi G, Murshudov GN, Luchinat C (2014) Simultaneous use of solution NMR and X-ray data in REFMAC5 for joint refinement/detection of structural differences. Acta Cryst D70:958–967

Chapter 22
Small and Wide-Angle Scattering Beamlines

SAXS/WAXS (X stands for X-ray) beamlines are usually available at all synchrotrons and XFEL facilities and are too numerous to summarise here. Likewise, SANS (N stands for neutrons) beamlines are usually available at all neutron facilities and are too numerous to summarise. The IUCr Commission on Small Angle Scattering is described at: https://www.iucr.org/resources/commissions/small-angle-scattering and the IUCr Commission on Neutron Scattering is described at: https://www.iucr.org/iucr/commissions/neutron-scattering.

Good practice outcomes of a round robin study for biological small angle scattering are described in Trewhella et al. (2022). Pauw et al (2023) describe a materials science round robin comprising four datasets of 1D scattering data, representative of two dilute and two dense nanoparticle systems, that were made available to the 46 participants. These two round robin studies are an important work towards the certification of the findings from these scattering methods. A comprehensive text describing condensed matter applications is that of Boothroyd (2020). SANS has special strengths in contrast matching applications.

Major textbooks are the monograph by Lattman et al (2018) and the teaching text of Svergun et al (2013). Both books emphasise the growth in popularity and importance of these solution scattering methods in structural biology. The most recent one, by Lattman et al (2018), emphasises the dramatic increase in the power of methods arising from "intense third-generation X-ray sources, low noise detectors, new algorithms and the computational power to take advantage of all of these have appeared in the past decade". They also point out that workshops on these methods are popular.

J. R. Helliwell, *Certifying Central Facility Beamlines for Biological and Chemical Crystallography and Allied Methods*,
SpringerBriefs in Crystallography, https://doi.org/10.1007/978-3-031-80181-5_22

References

Boothroyd AT (2020) Principles of neutron scattering from condensed matter. Oxford University Press, Oxford, p 512

Lattman EE, Grant TD, Snell EH (2018) Biological small angle scattering. IUCr monographs on crystallography. Oxford University Press, Oxford

Pauw BR, Smales GJ, Anker AS, Annadurai V, Balazs DM, Bienert R, Bouwman WG, Brebler I, Breternitz J, Brok ES, Bryant G, Clulow AJ, Crater ER, De Geuser F, Del Giudice A, Deumer J, Disch S, Dutt S, Frank K, Fratini E, Garcia PRAF, Gilbert EP, Hahn MB, Hallett J, Hohenschutz M, Hollamby M, Huband S, Ilavsky J, Jochum JK, Juelsholt M, Mansel BW, Penttilla P, Pittkowski RK, Portale G, Pozzo LD, Rochels L, Rosalie JM, Saloga PEJ, Seibt S, Smith AJ, Smith GN, Spiering GA, Stawski TM, Tache O, Thunemann AF, Toth K, Whitten AE, Wuttke J (2023) The human factor: results of a small angle scattering data analysis round robin. J Appl Cryst 56:1618–1629

Svergun DI, Koch MHJ, Timmins PA, May RP (2013) Small angle X-ray and neutron scattering from solutions of biological macromolecules. IUCr texts on crystallography. Oxford University Press, Oxford

Trewhella J, Vachette P, Bierma J, Blanchet C, Brookes E, Chakravarthy S, Chatzimagas L, Cleveland T, Cowieson N, Crossett B, Duff A, Franke D, Gabel F, Gillilan R, Graewert M, Grishaev A, Guss J, Hammel M, Hopkins J, Huang Q, Hub J, Hura G, Irving T, Jeffries C, Jeong C, Kirby N, Krueger S, Martel A, Matsui T, Li N, Perez J, Porcar L, Prange T, Rajkovic I, Rocco M, Rosenberg D, Ryan T, Seifert S, Sekiguchi H, Svergun D, MarujoTeixeira S, Thureau A, Weiss T, Whitten A, Wood K, Zuo X (2022) A round-robin approach provides a detailed assessment of biomolecular small-angle scattering data reproducibility and yields consensus curves for benchmarking. Acta Cryst D78:1315–1336

Chapter 23
X-Ray Absorption (XAS) Spectroscopy Beamlines

X-rays are absorbed by matter primarily through the photoelectric effect. Photoelectric absorption occurs when a bound electron (e.g. K shell) is excited to a continuum state by an incident photon of energy. It is thereby possible to discriminate between different elements in a given sample. By using not only K but also L edges and possibly M edges all elements in the periodic table are accessible using synchrotron radiation. Since the effect is dependent only on the presence of an excitable atom and an ordered local environment of that atom, the sample can be disordered (e.g. amorphous, solution) or ordered (e.g. crystalline); the only constraint being that sufficient material be present and the concentration of the excitable atom be enough for a reasonable signal-to-noise ratio. Information concerning the electronic structure and/or immediate environment about the primary absorbing atom can be obtained by the accurate measurement of: the position (point of maximum inflection) of the absorption edge, the details of the edge structure, and the intervals and amplitudes of the Extended X-ray Absorption Fine Structure (EXAFS) ripples. An energy dispersive detector can be used to isolate the fluorescent radiation from the Compton and Rayleigh components important for dilute systems.

The analysis of the data aims at extracting information on the local neighbours of the stimulated atom: namely the number, distance, and type of neighbours. EXAFS derived distances are more precise than those usually derived from macromolecular crystallography.

XAS beamlines are usually available at all synchrotrons and XFEL facilities and are too numerous to summarise here. A website summarising books and reviews on XAS is here https://xrayabsorption.org/books-and-review-articles/. The IUCr Commission on XAFS is described here https://www.iucr.org/resources/commissions/xafs.

© The Author(s), under exclusive license to Springer Nature Switzerland AG 2025   67
J. R. Helliwell, *Certifying Central Facility Beamlines for Biological and Chemical Crystallography and Allied Methods*,
SpringerBriefs in Crystallography, https://doi.org/10.1007/978-3-031-80181-5_23

Chapter 24
Mass Spectroscopy

Mass spectroscopy (also known as mass spectrometry) is an analytical technique that is used to measure the mass to charge ratio of ions with the measurements presented as a mass spectrum. It is used in many different fields and is applied to pure samples as well as complex mixtures. These instruments are provided commercially such as described here https://www.agilent.com/.

Ebsworth et al. (1991) give a detailed description of mass spectrometry methods and applications in inorganic chemistry. Mass spectrometry in structural biology has grown impressively in its power and scope these last decades (see e.g. Liko et al. 2016). Historically Bragg (1913) used the mass estimate to decide on the composition of the alkali halide crystals (see his page 272).

References

Bragg WL (1913) The structure of some crystals as indicated by their diffraction of X-rays Proc. R Soc London Ser A 89:248–277

Ebsworth EAV, Rankin DWH, Cradock S (1991) Structural methods in inorganic chemistry, 2nd ed. Blackwell, Oxford, 528p (Chapter 9 describes mass spectrometry)

Liko I, Allison TM, Hopper JTS, Robinson CV (2016) Mass spectrometry guided structural biology. Curr Opin Struct Biol 40:136–144

Chapter 25
UV/Vis and Infra-Red Spectroscopy

Fourier transform UV–Vis, (FT)–IR, Raman spectroscopies, in addition to NMR spectroscopy, are some of the most routinely used techniques in chemistry laboratories. These methods provide complementary information to crystallography but in solution. A comprehensive treatise on many different structural methods in inorganic chemistry, diffraction and spectroscopy based, is again the excellent book by Ebsworth et al (1991).

UV/Visible spectrophotometers are available in basically every chemistry department, be it academic or industrial. The sample is a solution placed in a measuring cell through which the UV/Vis light passes and the absorption as a function of light frequency or wavelength is measured. Solid state samples are somewhat more difficult to measure from but a diffuse reflectance addition to the spectrophotometer allows the reflected light spectrum to be analysed in the same manner as the transmitted light case for a solution. In the absence of the additional device then the solid sample, even a crystal if soft enough, can be squeezed i.e. crushed between two microscope slides to make a sample thin enough to transmit the UV/Vis light to the measuring photometer. An example of that approach is described in Bartalucci et al (2009) who describe both solution and crushed carotenoid crystals measurements: *"UV–vis spectra were measured on a Varian Cary 5000 UV–vis–NIR spectrometer. The solution-state spectra were recorded with the carotenoids dissolved in chloroform. The solid-state spectra were obtained from crystals pressed between two glass coverslips."*

Helliwell and Massera (2022) survey the roles of complementary methods which together yield accuracy in structural studies (i.e. each method is precise but together insights into accuracy can be gained). They give examples such as involving FT–IR and Raman spectroscopies; both involve the study of the interactions of radiation with the molecular vibrations of a sample and are generally associated with the bond strength between atoms in molecules. Moreover, they can help clarify problems that cannot be solved solely through X-ray diffraction.

© The Author(s), under exclusive license to Springer Nature Switzerland AG 2025   71
J. R. Helliwell, *Certifying Central Facility Beamlines for Biological and Chemical Crystallography and Allied Methods*,
SpringerBriefs in Crystallography, https://doi.org/10.1007/978-3-031-80181-5_25

Infrared spectroscopy offers a variety of possible measurements (see e.g. Barth (2007)). It can be applied in both static and time-resolved studies. These latter include taking the difference spectra between two structural states. Such difference spectra can be very informative and can confirm and complement what is seen with 3D structural probes.

References

Bartalucci G, Fisher S, Helliwell JR, Helliwell M, Liaaen-Jensen S, Warren JE, Wilkinson J (2009) X-ray crystal structures of diacetates of 6-s-cis and 6-s-trans astaxanthin and of 7,8-didehydroastaxanthin and 7,8,7′,8′-tetradehydroastaxanthin: comparison with free and protein-bound astaxanthins. Acta Cryst B65:238–247

Barth A (2007) Infrared spectroscopy of proteins. Biochim Biophys Acta 1767:1073–1101

Ebsworth EAV, Rankin DWH, Cradock S (1991) Structural methods in inorganic chemistry, 2nd ed. Blackwell, Oxford, 528p

Helliwell JR, Massera C (2022) The four Rs and crystal structure analysis: reliability, reproducibility, replicability and reusability. J Appl Cryst 55:1351–1358

Chapter 26
Data Analysis and Software

A highly effective way to judge the quality of a beamline's performance, whether by a user or by a beamline scientist, is the data analysis. Across the domains of crystallography, diffraction, scattering, and spectroscopy, as well as cryoEM and NMR, there are software analysis packages available. Within crystallography itself there are distinct sub-domains for small molecule and for macromolecule data analysis and structure determination and analysis.

In macromolecular crystallography the most well-known and extensively known software packages are CCP4 (Agirre et al. 2023), PHENIX (Liebschner et al. 2019) and SHELX (Sheldrick 2008). Molecular graphics is dominated by the COOT software (Emsley et al. 2010) and for publication figures are prepared using CCP4MG (McNicholas et al. 2011) or PyMOL (DeLano, http://www.pymol.org) or ChimeraX (https://www.rbvi.ucsf.edu/chimerax).

In small molecule crystallography the most well-known and extensively known software packages are again SHELX (Sheldrick 2008) and Olex (Dolamanov et al. 2003) and Olex2 (Puschmann and Dolomanov (2006) Molecular graphics is dominated by the Mercury software from the CCDC (Macrae et al. 2020).

For the analysis of non-crystalline and fibre diffraction patterns the CCP13 project was established in the 1990s to generate software which will be generally useful (Rajkumar et al. (2007)). Powder diffraction software is provided in CCP14 (Stephenson et al. 2006). In cryoEM a summary of the software and processing and analysis methods are described by (Baldwin et al. (2018)). An example of NMR analysis is by Gryk et al. (2010).

Some beamlines and central facilities have made available analysis pipelines as well as the raw i.e. experimental data themselves. These pipelines can usefully incorporate a parallel analysis of a raw data set by the different softwares available. These allow the user to evaluate the spread of outcomes from these analysis packages. In my experience if a sample is not so perfect e.g. in macromolecular crystallography the crystal sample is perhaps quite mosaic ('high mosaicity') or diffracts anisotropically then the available softwares can produce a range of processing outcomes.

© The Author(s), under exclusive license to Springer Nature Switzerland AG 2025

J. R. Helliwell, *Certifying Central Facility Beamlines for Biological and Chemical Crystallography and Allied Methods*,

SpringerBriefs in Crystallography, https://doi.org/10.1007/978-3-031-80181-5_26

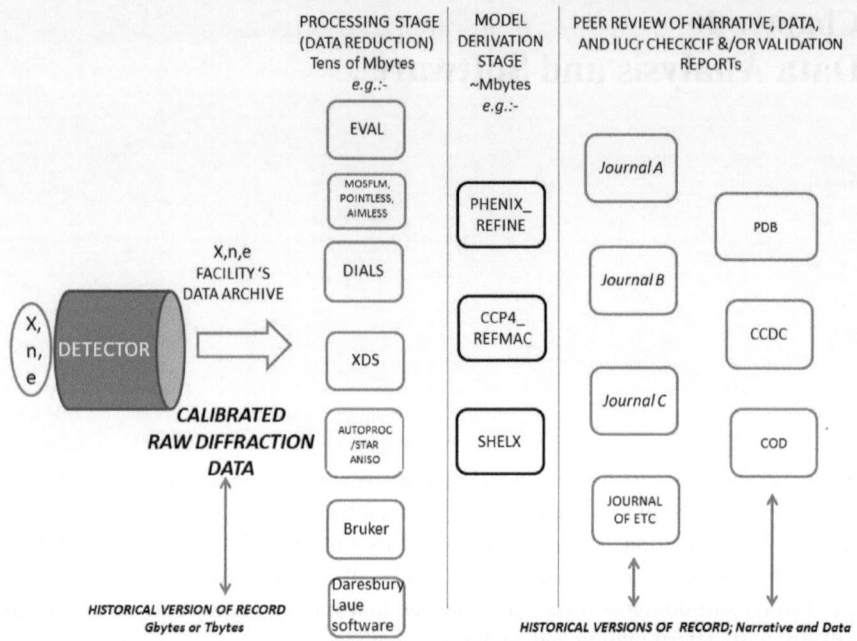

Fig. 26.1 The workflows in crystallography have a well organised set of stages, each one following standards agreed by the global crystallography community which thereby establish the provenance of a study. The references to each software package are: EVAL (Schreurs et al. 2010); iMOSFLM (Battye et al. 2011); AIMLESS and POINTLESS (Evans 2011); DIALS (Winter et al. 2018); XDS (Kabsch 2010a, 2010b); AUTOPROC (Vonrhein et al. 2011); STARANISO (Tickle et al. 2018); Bruker_SAINT (Bruker 2012); Daresbury Laue (Arzt et al. 1999; Hao et al. 2021); PHENIX_ REFINE (Liebschner et al. 2019); CCP4_REFMAC (Kovalevskiy et al. 2018; Agirre et al. 2023; neutron crystallographic refinement using REFMAC is described in Catapano et al. 2023); SHELX (Sheldrick 2008)

Even if the sample is perfect then choice of workflows to the final structure can produce slightly different molecular models. The differences seen using the different possible workflows of such as the B factors of atoms in a protein model can be useful as it allows an estimation of the variance on B values (for a recent discussion see (Helliwell (2023)). Figure 26.1 shows the possible choices of workflows from the experimental measurement stage through raw diffraction data processing to protein molecular model refinement then publication and database depositions. With artificial intelligence and machine learning approaches being increasingly available it can be imagined that whilst a human determines their preferred workflow, from such as their past experience, then AI/ML could deliver a better understanding of the variance of calculation outcomes such as from using different softwares and in differing combinations.

AI and ML software and usage is developing at a considerable pace. These can be for beamline setting up, their control and or calibration and even automatic protocols for making 'optimal' measurements at a beamline. Central facilities operate around the clock and so the use of AI and ML is attractive. This is not the same procedure as remote access (Smith et al. 2010) or telepresence (Warren et al. 2008); these offer convenience and reduced costs if a measuring team can operate from their home base and do not have to travel to the facility. But the use of AI and ML to 'replace' the human user during an experimental run is clearly different. A compilation of papers available on these topics, both for analysis as well as instrumentation and control has been brought together by Billinge and Proffen (2024). These topics are under very active discussion at crystallography conferences; e.g. there were several sessions at the ACA 2023 conference held in Baltimore USA (https://www.amercrystalassn. org/assets/Meeting/2022/PROSPECTUS%2023%20REVISED.pdf). This included sessions entitled: *2.2.2 Artificial Intelligence, Machine Learning, and Other Data Science Techniques Applied to Structure Determination, materials characterization, experiment control and data analysis* as well as *2.2.4 Machine learning in cryo-EM.*

I should close this section by confessing that this chapter is very hard to write as data analysis software and algorithms are constantly evolving. Also, the computer hardware is not static either. The new NVIDIA's GPU (circa March 2024) is described as "30 times faster than the H100 and use 25 times less power"; its impact in AI and ML is predicted to be a major leap.

References

Agirre J, Atanasova M, Bagdonas H, Ballard CB, Basle A, Beilsten-Edmands J, Borges RJ, Brown DG, Burgos-Marmol JJ, Berrisford JM, Bond PS, Caballero I, Catapano L, Chojnowski G, Cook AG, Cowtan KD, Croll TI, Debreczeni JE, Devenish NE, Dodson EJ, Drevon TR, Emsley P, Evans G, Evans PR, Fando M, Foadi J, Fuentes-Montero L, Garman EF, Gerstel M, Gildea RJ, Hatti K, Hekkelman ML, Heuser P, Hoh SW, Hough MA, Jenkins HT, Jimenez E, Joosten RP, Keegan RM, Keep N, Krissinel EB, Kolenko P, Kovalevskiy O, Lamzin VS, Lawson DM, Lebedev AA, Leslie AGW, Lohkamp B, Long F, Maly M, McCoy AJ, McNicholas SJ, Medina A, Millan C, Murray JW, Murshudov GN, Nicholls RA, Noble MEM, Oeffner R, Pannu NS, Parkhurst JM, Pearce N, Pereira J, Perrakis A, Powell HR, Read RJ, Rigden DJ, Rochira W, Sammito M, Sanchez Rodriguez F, Sheldrick GM, Shelley KL, Simkovic F, Simpkin AJ, Skubak P, Sobolev E, Steiner RA, Stevenson K, Tews I, Thomas JMH, Thorn A, Valls JT, Uski V, Usan I, Vagin A, Velankar S, Vollmar M, Walden H, Waterman D, Wilson KS, Winn MD, Winter G, Wojdyr M, Yamashita K (2023) The CCP4 suite: integrative software for macromolecular crystallography. Acta Cryst D79:449–461
Arzt S, Campbell JW, Harding MM, Hao Q, Helliwell JR (1999) LSCALE—the new normalisation, scaling and absorption correction program in the Daresbury Laue software suite. J Appl Cryst 32:554–562
Baldwin PR, Zi Tan Y, Eng ET, Rice WJ, Noble AJ, Negro CJ, Cianfrocco MA, Potter CS, Carragher B (2018) Big data in cryoEM: automated collection, processing and accessibility of EM data. Curr Opin Microbiol 43:1–8
Battye TGG, Kontogiannis L, Johnson O, Powell HR, Leslie AGW0 (2011) iMOSFLM: a new graphical interface for diffraction-image processing with MOSFLM. Acta Cryst D67:271–281
Billinge SJL, Proffen Th (2024) Machine learning in crystallography and structural science. Acta Cryst A80:139–145
Bruker (2012) SAINT Bruker AXS Inc., Madison, Wisconsin, USA

Catapano L, Long F, Yamashita K, Nicholls RA, Steiner RA, Murshudov GN (2023) Neutron crystallographic refinement with REFMAC5 from the CCP4 suite. Acta Cryst D79:1056–1070

Dolomanov OV, Blake AJ, Champness NR, Schroder M (2003) OLEX: new software for visualization and analysis of extended crystal structures. J Appl Cryst 36:1283–1284

Emsley P, Lohkamp B, Scott WG, Cowtan K (2010) Features and development of Coot. Acta Cryst D66:486–501

Evans PR (2011) An introduction to data reduction: space-group determination, scaling and intensity statistics. Acta Cryst D67:282–292

Gryk MR, Vyas J, Maciejewski MW (2010) Biomolecular NMR data analysis. Prog Nucl Magn Reson Spectrosc 56:329–345

Hao Q, Harding MM, Helliwell JR, Szebenyi DM (2021) Weblinks for the Daresbury Laue software source code and information. J Synchrotron Rad 28:666

Helliwell JR (2023) Error estimates in atom coordinates and B factors in macromolecular crystallography. Curr Res Struct Biol 6:100111

Kabsch W (2010a) XDS. Acta Cryst D66:125–132

Kabsch W (2010) Integration, scaling, space-group assignment and post-refinement. Acta Cryst D66:133–144

Kovalevskiy O, Nicholls RA, Long F, Carlon A, Murshudov GN (2018) Overview of refinement procedures within REFMAC5: utilizing data from different sources. Acta Cryst D74:215–227

Liebschner D, Afonine PV, Baker ML, Bunkczi G, Chen VB, Croll TI, Hintze B, Hung L.-W, Jain S, McCoy AJ, Moriarty NW, Oeffner RD, Poon BK, Prisant MG, Read RJ, Richardson JS, Richardson DC, Sammito MD, Sobolev OV, Stockwell DH, Terwilliger TC, Urzhumtsev AG, Videau LL, Williams CJ, Adams PD (2019) Macromolecular structure determination using Xrays, neutrons and electrons: recent developments in Phenix. Acta Cryst D75:861–877

McNicholas S, Potterton E, Wilson KS, Noble MEM (2011) Presenting your structures: the *CCP4mg* molecular-graphics software Acta Cryst D67:386–394

Macrae CF, Sovago I, Cottrell SJ, Galek PTA, McCabe P, Pidcock E, Platings M, Shields GP, Stevens JS, Towler M, Wood PA (2020) Mercury 4.0: from visualization to analysis, design and prediction. J Appl Cryst 53:226–235

Puschmann H, Dolomanov O (2006) Olex2: a comprehensive molecular graphics tool for small-molecule structures. Acta Cryst A 62:s246

Rajkumar G, AL-Khayat HA, Eakins F, Knupp C, Squire JM (2007) The CCP13 FibreFix program suite: semi-automated analysis of diffraction patterns from non-crystalline materials. J Appl Cryst 40:178–184

Schreurs AMM, Xian X, Kroon-Batenburg LMJ (2010) EVAL15: a diffraction data integration method based on ab initio predicted profiles. J Appl Cryst 43:70–82

Sheldrick GM (2008) A short history of SHELX. Acta Cryst A 64:112–122

Smith CA, Card GL, Cohen AE, Doukov TI, Eriksson T, Gonzalez AM, McPhillips SE, Dunten PW, Mathews II, Song J, Soltis SM (2010) Remote access to crystallography beamlines at SSRL: novel tools for training, education and collaboration. J Appl Cryst 43:1261–1270

Stephenson R, Cockcroft J, Watkin D, Cernik B (2006) The CCP14: freely available crystallographic software for academia (collaborative computational project number 14). Acta Cryst A 62:s251

Tickle IJ, Flensburg C, Keller P, Paciorek W, Sharff A, Vonrhein C, Bricogne G (2018) STARANISO. Global Phasing Ltd., Cambridge

Vonrhein C, Flensburg C, Keller P, Sharff A, Smart O, Paciorek W, Womack T, Bricogne G (2011) Data processing and analysis with the autoPROC toolbox. Acta Cryst D67:293–302

Warren JE, Diakun G, Bushnell-Wye G, Fisher S, Thalal A, Helliwell M, Helliwell JR (2008) Science experiments via telepresence at a synchrotron radiation source facility. J Synchrotron Rad 15:191–194

Winter G, Waterman DG, Parkhurst JM, Brewster AS, Gildea RJ, Gerstel M, FuentesMontero L, Vollmar M, Michels-Clark T, Young ID, Sauter NK, Evans G (2018) DIALS: implementation and evaluation of a new integration package. Acta Cryst D 74:85–97

Chapter 27
Safety Matters

In readiness for an experiment, it is vital to plan the safety of everything to do with it. This includes pre-beamtime training which will be provided by the facility. These are in my experience online and so can be viewed at one's convenience but must be completed before arriving at the facility's site. Sometimes the conclusion of viewing a video will be a quiz. If you fail one of the questions you may well have to repeat your viewing and the quiz.

Sample preparation will have to follow local safety procedures such as the Control of Substances Hazardous to Health (COSSH) regulations in the UK for chemistry. This will include the chemical hazards and the procedures to be used in their manipulation. These all need to be considered before the experiment and approved i.e. signed by the Team Leader.

It is very important to carefully track sample shipments and data collections, and laboratory information management systems (LIMS) have come to the fore like ISPYB (Fisher et al 2015). Likewise, there are the sample management information systems (SMIS) for beamtime applications and proposal management. These also connect to the safety of the samples on site.

In summary, synchrotron radiation is dangerous, and experiments are conducted in shielded hutches to minimize radiation exposure. Users conducting experiments in person, i.e. not remote access, will be required to go through a training course on safety as a user and will be tested on the safe operation of the hutch. If biochemical or chemical manipulation is to occur at the facility, there are also facility-specific safety regulations that must be adhered to. It is in the interests of the facility and the user to flag any potentially hazardous experiment well before the beamtime so that suitable precautions can be taken. Facility beamtime proposals require extensive details of the samples to be brought on site and their possible hazards.

© The Author(s), under exclusive license to Springer Nature Switzerland AG 2025   77
J. R. Helliwell, *Certifying Central Facility Beamlines for Biological and Chemical Crystallography and Allied Methods*,
SpringerBriefs in Crystallography, https://doi.org/10.1007/978-3-031-80181-5_27

Reference

Fisher SJ, Levik KE, Williams MA, Ashton AW, McAuley KE (2015) *SynchWeb*: a modern interface for *ISPyB*. J Appl Cryst 48:927–932

Chapter 28
Theoretical and Computational Sciences

All experiments are preceded with some level of theory, even if it is a basic question like "what happens if we do this?". This is referred to as "science pull". The wider context of an individual scientist is the degree of sophistication of the technology available. Advances in technology create a push to what scientists can think of doing, referred to as "technology push".

Scientists at the most fundamental level are seekers after truth. But truth is in essence unattainable experimentally due to systematic errors in our methods and random errors in our measurements. Combining different methods lets us know the accuracy of our findings. In each method we seek to reduce its systematic errors of measurement to a minimum and make enough measurements with it to get the best precision realistically possible. Theory can help again as it can identify the "ideal" achievable.

As a personal example I offer:-

> In a simple case of what is the best geometric perfection possible of a protein crystal I considered the perfect protein crystal as a theoretical construct to ask myself: what would its mosaicity be? (Helliwell 1988). This told me what a beamline source and optics would have to be like to reach that level of measurement; this took me to the ADONE synchrotron and its diffractometer on an unfocussed beamline in Frascati, Italy (Colapietro et al 1992)

In structural crystallography our studies of an enzyme of course must be done with a crystal, but the crystal growth conditions may take us well away from the pH where the enzyme functions in life. The enzyme's ionisable amino acids may then have protonation states that are 'false'. We have here what is sometimes called "approximate truth". The method of microgravity protein crystal growth gives a convection free and sedimentation free fluid state (see e.g. Snell and Helliwell 2021). So, the very tiny crystals that can grow on the ground at the enzyme's functional pH have every chance to grow larger in microgravity for X-ray or neutron crystallography studies on their return to earth.

Another strong feature of research in protein crystallography seeks to discover new compounds that bind to a protein binding site, i.e. a protein from a bacterial

J. R. Helliwell, *Certifying Central Facility Beamlines for Biological and Chemical Crystallography and Allied Methods*,
SpringerBriefs in Crystallography, https://doi.org/10.1007/978-3-031-80181-5_28

infecting organism or a viral spike protein. Such a compound might be developed into a drug. Computational screening methods seek to use a protein 3D structure and a library of potential ligands with their 3D structures to find ones that will likely bind to the protein. Bradbrook et al. (1998) investigated in a test case whether the thermodynamic calorimetry data of binding of two similar saccharide ligands to the lectin concanavalin A could be understood in terms of their 3D structures. The Gibbs free energy change difference between these two saccharides (mannoside and glucoside) on binding to the protein estimated by Bradbrook et al. (1998) could not be predicted with confidence based on the two protein crystal structures. However, when they added molecular dynamics simulations for each protein structure it allowed them to take account of transient hydrogen bonds of each saccharide to the protein. This gave better estimates of the Gibbs free energies. Meanwhile a totally different approach of "fragment screening" (Blundell 2017) provides an overview) yielded lead compounds experimentally. At various synchrotron radiation sources fragment screening beamlines now exist. These fixed wavelength beamlines can deliver thousands of new protein–ligand soaked crystal structures per day! See e.g. https://www.diamond.ac.uk/Instruments/Mx/Fragment-Screening.html. A caveat I would mention on this clearly important experimental method is that for the robotic sample changing needed to achieve the high throughput, frozen protein crystals are used. This approach means that artefactual binding sites may be seen, especially those at low occupancy, which will not be bound to the protein at physiologically relevant temperatures (Halle 2004). In situ macromolecular crystallography beamlines are now online so as to yield structures at room temperature (for a short interview re the Diamond Light Source VMXi beamline by Dr Mike Hough see https://www.jove.com/v/65964/author-spotlight-advancing-protein-structure-analysis-for-drug).

Data analysis has been covered in several other sections in this Springer Brief. It is a very important topic under the title of this chapter as well. In the interpretation of solution scattering measurements, the prediction of those scattering profiles for different macromolecular shapes as hypotheses can ensure that a SAXS experiment could distinguish them. In studies of enzyme reactions in crystals it would be nice to predict enzymatic pathways for time-resolved studies so that experimental time points could be predicted and tested. Alas, to my knowledge it is not yet possible, and is deemed a grand challenge to predict enzyme reaction rates for an isolated enzyme molecule let alone packed in a crystal; for a lucid discussion see Blow (2000). Manipulation of an enzyme reaction rate in a relative way is possible though such as amino acid mutation; for an example of a K59Q mutation in slowing down the reaction of hydroxymethylbilane synthase see Niemann et al (1994).

References

Blow DM (2000) So do we understand how enzymes work? Structure 8:R77–R81

Blundell TL (2017) Protein crystallography and drug discovery: recollections of knowledge exchange between academia and industry. IUCrJ 4:308–321

Bradbrook GM, Gleichmann T, Harrop SJ, Habash J, Raftery J, Kalb J, Yariv J, Hillier IH, Helliwell JR (1998) X-Ray and molecular dynamics studies of concanavalin-A glucoside and mannoside complexes Relating structure to thermodynamics of binding. J Chem Soc, Faraday Trans 94(11):1603–1611

Colapietro M, Cappuccio G, Marciante C, Pifferi A, Spagna R, Helliwell JR (1992) The X–ray diffraction station at the ADONE Wiggler facility: hardware, software, and preliminary results (including crystal perfection). J Appl Cryst 25:192–194

Halle B (2004) Protein hydration dynamics in solution: a critical survey. Proc Natl Acad Sci USA 101:4793–4798

Helliwell JR (1988) (1988) Protein crystal perfection and the nature of radiation damage. J. Crystal Growth 90:259–272

Niemann AC, Hädener A, Matzinger PK (1994) A kinetic analysis of the reaction catalysed by hydroxymethylbilane synthase. HCA 77:1791–1809. https://doi.org/10.1002/hlca.199407 70711

Snell EH, Helliwell JR (2021) Microgravity as an environment for macromolecular crystallization— an outlook in the era of space stations and commercial space flight. Crystallogr Rev 27(1):3–46

Chapter 29
Summary

There are two perspectives that are special to an experimental run at a synchrotron, XFEL or neutron facility: that of the user and that of the facility staff. The user looks to the website for their beamline of interest as a first step to their certification of it i.e. that it will do what they envisage. The facility staff trust that the user has a scientifically excellent experiment, having been approved for beamtime by their beamtime approval committee. The user also must fully commit to the beamtime scheduled. The facility also does its best to make clear their policies for use. A key step is the Principal Investigator signing up to the "tick boxes" at the time of the proposal. These tick boxes are such as:

- agreeing the "Data management and sharing plan";
- declaring the details of safety of the sample and the apparatus that may be brought on site as well as the beamline;
- agreeing the publication authorship or acknowledgements guidelines/regulations in the facility's policy.

Some experiments require multiple beamlines or may sometimes require multiple probes i.e. X-ray, neutron, electron and/or NMR facilities. These experiments are the most challenging experimental proposals administratively, as one facility's use may be contingent on success at the other(s) and yet timeliness of gathering all the data from multiple probes requires in effect parallel measurements not sequential beamtime runs.

Some research themes such as battery research require huge teams knowing what each team member is planning or doing. At our crystallography conferences and congresses, we may have 10% of all the keynote lectures and 10% of the micro symposia devoted to battery research. Such is the interest in, and societal need for, better battery performance. These form a fairly new category of research at beamlines in the landscape of today.

J. R. Helliwell, *Certifying Central Facility Beamlines for Biological and Chemical Crystallography and Allied Methods*,
SpringerBriefs in Crystallography, https://doi.org/10.1007/978-3-031-80181-5_29

Chapter 30
So, to the Measuring Team Briefing Before the Experiment

The Facility has scheduled your experiment. Its staff will have worked hard to ensure the smooth running of its beamline and instruments for your experiment. So, onto the briefing. An onsite experiment tends to be more complicated than a remote experiment. The briefing will likely involve the Principal Investigator (PI) leading a discussion with the measuring team leader (a senior PostDoc let's say) and Ph.D. students. If I may use a military analogy the PI as Commander will describe the overall strategy and the measuring team leader is the General on the ground, with the troops, and who will decide tactics commensurate with achieving the overall strategy. The beamline or instrument selected will have been according to the themes described in all the chapters of this book. As just one example, at the beamline, the measuring team will have to settle detailed questions of: the desired diffraction resolution, the size and stability of the available (possibly very small) crystals, ensuring that at least a fluorescence scan is made as well to confirm presence of particular metal(s) in the sample, and or the possibility of some crystal sample phase transitions or unexpected changes of the crystal's space group.

Let's consider the overall strategy in some depth and breadth, which the pre-experiment briefing can usefully be reviewed. Experiments require a plan, made obviously by the PI for each project as a beamtime proposal to a Facility. At this strategic plan level, for the molecule being investigated, the aim may be location of a ligand at a reaction site and its interactions, possibly mediated by bound water molecules. These latter questions may then lead to the synchrotron only experiment becoming a joint refinement of synchrotron and neutron data or, in the future, a triply joint refinement adding in electron data or indeed a quadruple joint refinement if NMR data are also added in. Stepping up to the most general level: what results will be most useful to facilitate the basic science discovery or applied science impact desired? In either category, as an example, the need for biologically relevant temperature protein structures suggest that we should not settle for room temperature as necessarily physiological. Medical driven research looking for ligands as drug molecules for a human patient strongly suggests a check of the body temperature (37 °C) crystal structure;

© The Author(s), under exclusive license to Springer Nature Switzerland AG 2025

J. R. Helliwell, *Certifying Central Facility Beamlines for Biological and Chemical Crystallography and Allied Methods*,

SpringerBriefs in Crystallography, https://doi.org/10.1007/978-3-031-80181-5_30

see Jacobs et al (2024). In turn the wish for the crystallisation to be at 37 °C as well might be desirable, but difficult if the molecule in question is then too soluble to be crystallised. Even more challenging on that point would be seeking temperature relevant crystal structures for thermophiles or hyperthermophiles to understand their adaptation to such high temperature environments. The same would apply to understanding the molecular basis of survival in other extreme environments such as high salt. I commend the book by Wharton (2002) entitled *Life at the Limits*. So, we see in this elaboration at the overall scientifically desirable down to the very basic levels of detailed measurement choices, compromises of the crystallisation or experimental protocol may well be needed.

In another direction of research, we may seek the hydrogenation details of an enzyme catalytic site and whereby the pH of the natural living state is quite different from the pH where the molecule will crystallize. However, it is clearly the case that ever smaller crystals have become viable for diffraction data collection be this at synchrotrons or XFELS for X-rays or neutron sources with their modern instruments. So, crystallisation at the scientifically desirable pH comes closer.

The PI of a macromolecular crystallography project, in submitting a funding proposal, may well have had to deal with a referee's comment: *'why bother doing protein crystal structures when Alphafold exists?'*. The reply by the PI will have been that the prediction of the fold is one thing, and it may well be successful in the crystallography initial phase determination, but a prediction is exactly that. Molecular structure details matter and should be established by experiment.

In chemical crystallography the most basic anxiety, rarely expressed today in my experience, is the extent to which crystal packing might amend a molecule and its structure. In times long past there was an enthusiasm for gas phase electron diffraction and molecular structure elucidation from those data. Leading crystallographers of the past decades were practitioners of gas phase electron diffraction such as Jerry Karle (Nobel Prize in Chemistry 1985) and Durward Cruickshank. For a description of the past 50 years of gas phase electron diffraction up to and including a modern-day example see Wagner et al. (2021).

The Measuring Team will have a range of years of experience of team members as scientists, and in our area of research, a range of trainings by scientific subject. At risk of being pedantic, but for the avoidance of doubt, I mention some other basic points of a more nomenclature type. Firstly, our diffraction experiment, being a single method, in a physicist's terminology delivers a certain level of precision and not accuracy. Any experimental method has its own systematic errors, which are minimised, and so two or more methods' results need to be compared to assess accuracy. On a practical detail, as Cruickshank (1999) emphasised with his Diffraction Precision Index formulae, the completeness of the diffraction data impacts on the precision. Neutron Laue crystallography in particular must strive for as high a completeness as possible even if the measuring run takes a week or two rather than a couple of days.

Also on nomenclature, studies aiming at determining function of a molecule via time-resolved structure from diffraction measurements, will be under the overall heading of *structural dynamics*. I have never really liked this term, always preferring the *'time-resolved'* label. Why? Structural dynamics is an ambiguous label as it includes: "what the B factors are?" which quantify the atomic vibrations and/or static disorder; movements of atoms or not; appearance of atoms such as atoms becoming ordered or a structural intermediate appearing after addition of a reactant; disappearance of atoms meaning they have become disordered or removed from the molecule of interest. The measuring team, and the subsequent analytical team, need to be on the lookout for all these aspects of structural dynamics, and making a proper assessment of the model uncertainties on each of them.

As a final sentence, I hope this Brief book proves useful to readers.

References

Cruickshank DWJ (1999) Remarks about protein structure precision. Acta Cryst D55:583–601

Jacobs FJF, Helliwell JR, Brink A (2024) Body temperature protein X-ray crystallography at 37°C: a rhenium protein complex seeking a physiological condition structure. Chem Comm 60(95):14030–14033

Wagner I, Rankin DWH, Masters SL (2021) Gas electron diffraction then and now: from trisilyl phosphine to iso-propyl(*tert*-butyl)(trichlorosilyl)phosphine. Dalton Trans 50:17000–17007

Wharton DA (2002) Life at the limits: organisms in extreme environments. Cambridge University Press, Cambridge

Correction to: Certifying Central Facility Beamlines for Biological and Chemical Crystallography and Allied Methods

Correction to:
J. R. Helliwell, *Certifying Central Facility Beamlines*
for Biological and Chemical Crystallography and Allied
Methods, **SpringerBriefs in Crystallography,**
https://doi.org/10.1007/978-3-031-80181-5

The original version of the book was updated with the belated corrections in book chapter 16 and 17. The book and the chapters have been updated.

The updated versions of these chapters can be found at
https://doi.org/10.1007/978-3-031-80181-5_16
https://doi.org/10.1007/978-3-031-80181-5_17

References

Helliwell JR (1992) Global instrumentation survey: macromolecular crystallography. Synchrotron Radiation News 5(2):22–26. https://doi.org/10.1080/08940889208602669. This survey provided an overview of the performances of the user available beamlines for macromolecular crystallography at that time

Helliwell JR (1992) Macromolecular crystallography with synchrotron radiation. Cambridge University Press, Cambridge. Published in paperback 2005. Section 5.6 *"Gazetteer of SR workstations for macromolecular crystallography"*. Largely of historical interest it nevertheless provides a snapshot of facilities commitment to macromolecular crystallography up to that time

Helliwell JR (2021) Combining X-rays, neutrons and electrons, and NMR, for precision and accuracy in structure–function studies Acta Cryst A77:173–185

The Journal of Video Education https://www.jove.com/ has a very good range of videos and accompanying articles for X-ray, neutron, and electron crystallography which I have cited in the main text sections. There are plenty more, again of high quality, with each video accompanied by an explanatory article. They describe themselves as follows *"Filmed at the world's top scientific institutions, JoVE videos bring to life the intricate details of cutting-edge experiments enabling efficient learning and replication of new research methods and technologies."* They each explain the measuring and data analysis protocols

J. R. Helliwell, *Certifying Central Facility Beamlines for Biological and Chemical Crystallography and Allied Methods*,
SpringerBriefs in Crystallography, https://doi.org/10.1007/978-3-031-80181-5

Subject Index